Spatial Data Science

Spatial Data Science introduces fundamental aspects of spatial data that every data scientist should know before they start working with spatial data. These aspects include how geometries are represented, coordinate reference systems (projections, datums), the fact that the Earth is round and its consequences for analysis, and how attributes of geometries can relate to geometries. In the second part of the book, these concepts are illustrated with data science examples using the R language. In the third part, statistical modelling approaches are demonstrated using real world data examples. After reading this book, a number of major spatial data analysis errors should no longer be made because of lack of knowledge.

The book gives a detailed explanation of the core spatial software packages for R: **sf** for simple feature access, and **stars** for raster and vector data cubes – array data with spatial and temporal dimensions. It also shows how geometrical operations change when going from a flat space to the surface of a sphere, which is what **sf** and **stars** use when coordinates are not projected (degrees longitude/latitude). Separate chapters detail a variety of plotting approaches for spatial maps using R, and different ways of handling very large vector or raster (imagery) datasets, locally, in databases, or in the cloud. The data used and all code examples are freely available online from https://r-spatial.org/book/. The solutions to the exercises can be found here: https://edzer.github.io/sdsr_exercises/.

Chapman & Hall/CRC
The R Series

Series Editors
John M. Chambers, Department of Statistics, Stanford University, California, USA
Torsten Hothorn, Division of Biostatistics, University of Zurich, Switzerland
Duncan Temple Lang, Department of Statistics, University of California, Davis, USA
Hadley Wickham, RStudio, Boston, Massachusetts, USA

Recently Published Titles

Advanced R Solutions
Malte Grosser, Henning Bumann, and Hadley Wickham

Event History Analysis with R, Second Edition
Göran Broström

Behavior Analysis with Machine Learning Using R
Enrique Garcia Ceja

Rasch Measurement Theory Analysis in R: Illustrations and Practical Guidance for Researchers and Practitioners
Stefanie Wind and Cheng Hua

Spatial Sampling with R
Dick R. Brus

Crime by the Numbers: A Criminologist's Guide to R
Jacob Kaplan

Analyzing US Census Data: Methods, Maps, and Models in R
Kyle Walker

ANOVA and Mixed Models: A Short Introduction Using R
Lukas Meier

Tidy Finance with R
Stefan Voigt, Patrick Weiss and Christoph Scheuch

Deep Learning and Scientific Computing with R torch
Sigrid Keydana

Model-Based Clustering, Classification, and Density Estimation Using mclust in R
Lucca Scrucca, Chris Fraley, T. Brendan Murphy, and Adrian E. Raftery

Spatial Data Science: With Applications in R
Edzer Pebesma and Roger Bivand

For more information about this series, please visit: https://www.crcpress.com/Chapman--HallCRC-The-R-Series/book-series/CRCTHERSER

Spatial Data Science
With Applications in R

Edzer Pebesma
Roger Bivand

CRC Press
Taylor & Francis Group
Boca Raton London New York

CRC Press is an imprint of the
Taylor & Francis Group, an **informa** business

A CHAPMAN & HALL BOOK

Cover artwork by Allison Horst

First edition published 2023
by CRC Press
6000 Broken Sound Parkway NW, Suite 300, Boca Raton, FL 33487-2742

and by CRC Press
4 Park Square, Milton Park, Abingdon, Oxon, OX14 4RN

CRC Press is an imprint of Taylor & Francis Group, LLC

ISBN: 978-1-138-31118-3 (hbk)
ISBN: 978-1-032-47392-5 (pbk)
ISBN: 978-0-429-45901-6 (ebk)

DOI: 10.1201/9780429459016

Typeset in Latin Modern font
by KnowledgeWorks Global Ltd.

Publisher's note: This book has been prepared from camera-ready copy provided by the authors.

Access the Support Material: https://r-spatial.org/book/.

Table of contents

Preface   xi

I Spatial Data   1

1 Getting Started   5
  1.1 A first map   5
  1.2 Coordinate reference systems   7
  1.3 Raster and vector data   8
  1.4 Raster types   10
  1.5 Time series, arrays, data cubes   11
  1.6 Support   12
  1.7 Spatial data science software   13
      1.7.1 GDAL   14
      1.7.2 PROJ   14
      1.7.3 GEOS and s2geometry   14
      1.7.4 NetCDF, udunits2, liblwgeom   14
  1.8 Exercises   15

2 Coordinates   17
  2.1 Quantities, units, datum   17
  2.2 Ellipsoidal coordinates   18
      2.2.1 Spherical or ellipsoidal coordinates   19
      2.2.2 Projected coordinates, distances   21
      2.2.3 Bounded and unbounded spaces   22
  2.3 Coordinate reference systems   22
  2.4 PROJ and mapping accuracy   23
  2.5 WKT-2   26
  2.6 Exercises   27

3 Geometries   29
  3.1 Simple feature geometries   29
      3.1.1 The big seven   29
      3.1.2 Simple and valid geometries, ring direction   31
      3.1.3 Z and M coordinates   31
      3.1.4 Empty geometries   32
      3.1.5 Ten further geometry types   32
      3.1.6 Text and binary encodings   33

3.2 Operations on geometries 34
    3.2.1 Unary predicates 34
    3.2.2 Binary predicates and DE-9IM 34
    3.2.3 Unary measures 36
    3.2.4 Binary measures 37
    3.2.5 Unary transformers 37
    3.2.6 Binary transformers 38
    3.2.7 N-ary transformers 38
3.3 Precision 39
3.4 Coverages: tessellations and rasters 40
    3.4.1 Topological models 40
    3.4.2 Raster tessellations 41
3.5 Networks 42
3.6 Exercises 42

4 Spherical Geometries   **45**
4.1 Straight lines 45
4.2 Ring direction and full polygon 45
4.3 Bounding box, rectangle, and cap 46
4.4 Validity on the sphere 47
4.5 Exercises 47

5 Attributes and Support   **49**
5.1 Attribute-geometry relationships and support 50
5.2 Aggregating and summarising 52
5.3 Area-weighted interpolation 54
    5.3.1 Spatially extensive and intensive variables 54
    5.3.2 Dasymetric mapping 55
    5.3.3 Support in file formats 55
5.4 Up- and Downscaling 56
5.5 Exercises 57

6 Data Cubes   **59**
6.1 A four-dimensional data cube 59
6.2 Dimensions, attributes, and support 60
    6.2.1 Regular dimensions, GDAL's geotransform 62
    6.2.2 Support along cube dimensions 62
6.3 Operations on data cubes 63
    6.3.1 Slicing a cube: filter 63
    6.3.2 Applying functions to dimensions 64
    6.3.3 Reducing dimensions 64
6.4 Aggregating raster to vector cubes 65
6.5 Switching dimension with attributes 67
6.6 Other dynamic spatial data 67
6.7 Exercises 68

II R for Spatial Data Science   71

7 Introduction to sf and stars   **75**
  7.1 Package **sf** 75
    7.1.1 Creation 76
    7.1.2 Reading and writing 77
    7.1.3 Subsetting 78
    7.1.4 Binary predicates 78
    7.1.5 tidyverse 80
  7.2 Spatial joins 81
    7.2.1 Sampling, gridding, interpolating 82
  7.3 Ellipsoidal coordinates 82
  7.4 Package **stars** 84
    7.4.1 Reading and writing raster data 85
    7.4.2 Subsetting **stars** data cubes 86
    7.4.3 Cropping 88
    7.4.4 Redimensioning and combining **stars** objects 89
    7.4.5 Extracting point samples, aggregating 91
    7.4.6 Predictive models 92
    7.4.7 Plotting raster data 93
    7.4.8 Analysing raster data 93
    7.4.9 Curvilinear rasters 98
    7.4.10 GDAL utils 98
  7.5 Vector data cube examples 99
    7.5.1 Example: aggregating air quality time series 99
    7.5.2 Example: Bristol origin-destination data cube 102
    7.5.3 Tidy array data 108
    7.5.4 File formats for vector data cubes 109
  7.6 Raster-to-vector, vector-to-raster 109
    7.6.1 Vector-to-raster 109
  7.7 Coordinate transformations and conversions 110
    7.7.1 `st_crs` 110
    7.7.2 `st_transform`, `sf_project` 111
    7.7.3 `sf_proj_info` 112
    7.7.4 Datum grids, proj.db, cdn.proj.org, local cache 112
    7.7.5 Transformation pipelines 113
    7.7.6 Axis order and direction 115
  7.8 Transforming and warping rasters 116
  7.9 Exercises 117

8 Plotting spatial data   **119**
  8.1 Every plot is a projection 119
    8.1.1 What is a good projection for my data? 120
  8.2 Plotting points, lines, polygons, grid cells 121
    8.2.1 Colours 121

| 8.2.2 | Colour breaks: `classInt` . . . . | 122 |

    8.2.2   Colour breaks: `classInt` 122
    8.2.3   Graticule and other navigation aids 123
  8.3 Base `plot` 123
    8.3.1   Adding to plots with legends 123
    8.3.2   Projections in base plots 125
    8.3.3   Colours and colour breaks 125
  8.4 Maps with `ggplot2` 125
  8.5 Maps with `tmap` 127
  8.6 Interactive maps: `leaflet`, `mapview`, `tmap` 128
  8.7 Exercises 129

9 Large data and cloud native   **131**
  9.1 Vector data: `sf` 132
    9.1.1   Reading from local disk 132
    9.1.2   Reading from databases, **dbplyr** 133
    9.1.3   Reading from online resources or web services 134
    9.1.4   APIs, OpenStreetMap 134
    9.1.5   GeoParquet and GeoArrow 135
  9.2 Raster data: **stars** 136
    9.2.1   `stars` proxy objects 137
    9.2.2   Operations on proxy objects 139
    9.2.3   Remote raster resources 139
  9.3 Very large data cubes 140
    9.3.1   Finding and processing assets 140
    9.3.2   Cloud native storage: Zarr 140
    9.3.3   APIs for data: GEE, openEO 141
  9.4 Exercises 142

III   Models for Spatial Data   **143**

10 Statistical modelling of spatial data   **147**
  10.1 Mapping with non-spatial regression and ML models 148
  10.2 Support and statistical modelling 149
  10.3 Time in predictive models 150
  10.4 Design-based and model-based inference 151
  10.5 Predictive models with coordinates 152
  10.6 Exercises 153

11 Point Pattern Analysis   **155**
  11.1 Observation window 156
  11.2 Coordinate reference systems 160
  11.3 Marked point patterns, points on linear networks 161
  11.4 Spatial sampling and simulating a point process 163
  11.5 Simulating points on the sphere 164
  11.6 Exercises 164

12 Spatial Interpolation   **165**
     12.1 A first dataset 165
     12.2 Sample variogram 167
     12.3 Fitting variogram models 169
     12.4 Kriging interpolation 171
     12.5 Areal means: block kriging 171
     12.6 Conditional simulation 173
     12.7 Trend models 174
          12.7.1 A population grid 175
     12.8 Exercises 179

13 Multivariate and Spatiotemporal Geostatistics   **181**
     13.1 Preparing the air quality dataset 181
     13.2 Multivariable geostatistics 183
     13.3 Spatiotemporal geostatistics 184
          13.3.1 A spatiotemporal variogram model 184
          13.3.2 Irregular space time data 189
     13.4 Exercises 190

14 Proximity and Areal Data   **191**
     14.1 Representing proximity in `spdep` 192
     14.2 Contiguous neighbours 194
     14.3 Graph-based neighbours 197
     14.4 Distance-based neighbours 199
     14.5 Weights specification 204
     14.6 Higher order neighbours 206
     14.7 Exercises 208

15 Measures of Spatial Autocorrelation   **209**
     15.1 Measures and process misspecification 209
     15.2 Global measures 211
          15.2.1 Join-count tests for categorical data 212
          15.2.2 Moran's I 214
     15.3 Local measures 216
          15.3.1 Local Moran's I_i 218
          15.3.2 Local Getis-Ord G_i 225
          15.3.3 Local Geary's C_i 226
          15.3.4 The **rgeoda** package 229
     15.4 Exercises 231

16 Spatial Regression   **233**
     16.1 Markov random field and multilevel models 233
          16.1.1 Boston house value dataset 235
     16.2 Multilevel models of the Boston dataset 237
          16.2.1 IID random effects with lme4 238
          16.2.2 IID and CAR random effects with hglm 238

16.2.3 IID and ICAR random effects with R2BayesX   239
16.2.4 IID, ICAR and Leroux random effects with INLA . .   240
16.2.5 ICAR random effects with mgcv::gam() 241
16.2.6 Upper-level random effects: summary 241
16.3 Exercises   243

17 Spatial Econometrics Models   **245**
17.1 Spatial econometric models: definitions   245
17.2 Maximum likelihood estimation in **spatialreg**   248
17.2.1 Boston house value dataset examples   249
17.3 Impacts   252
17.4 Predictions   254
17.5 Exercises   256

A Older R Spatial Packages   **259**
A.1 Retiring **rgdal** and **rgeos**   259
A.2 Links and differences between sf and sp   259
A.3 Migration code and packages   260
A.4 Package raster and terra   260

B R Basics   **263**
B.1 Pipes   263
B.2 Data structures   264
B.2.1 Homogeneous vectors   264
B.2.2 Heterogeneous vectors: `list`   265
B.2.3 NULL and removing list elements 267
B.2.4 Attributes 267
B.2.5 The `names` attributes   270
B.2.6 Using `structure`   271
B.3 Dissecting a `MULTIPOLYGON`   272

References   **277**

Index   **291**

Index of functions   **299**

Preface

Data science is concerned with finding answers to questions on the basis of available data, and communicating that effort. Besides showing the results, this communication involves sharing the data used, but also exposing the path that led to the answers in a comprehensive and reproducible way. It also acknowledges the fact that available data may not be sufficient to answer questions, and that any answers are conditional on the data collection or sampling protocols employed.

This book introduces and explains the concepts underlying *spatial* data: points, lines, polygons, rasters, coverages, geometry attributes, data cubes, reference systems, as well as higher-level concepts including how attributes relate to geometries and how this affects analysis. The relationship of attributes to geometries is known as support, and changing support also changes the characteristics of attributes. Some data generation processes are continuous in space, and may be observed everywhere. Others are discrete, observed in tesselated containers. In modern spatial data analysis, tesellated methods are often used for all data, extending across the legacy partition into point process, geostatistical and lattice models. It is support (and the understanding of support) that underlies the importance of spatial representation. The book aims at data scientists who want to get a grip on using spatial data in their analysis. To exemplify how to do things, it uses R. In future editions we hope to extend this with examples using Python (see, e.g., Bivand 2022b) and Julia.

It is often thought that spatial data boils down to having observations' longitude and latitude in a dataset, and treating these just like any other variable. This carries the risk of missed opportunities and meaningless analyses. For instance,

- coordinate pairs really are pairs, and lose much of their meaning when treated independently
- rather than having point locations, observations are often associated with spatial lines, areas, or grid cells
- spatial distances between observations are often not well represented by straight-line distances, but by great circle distances, distances through networks, or by measuring the effort it takes getting from A to B

We introduce the concepts behind spatial data, coordinate reference systems, spatial analysis, and introduce a number of packages, including **sf** (Pebesma 2018, 2022b), **stars** (Pebesma 2022d), **s2** (Dunnington, Pebesma, and Rubak

2023) and **lwgeom** (Pebesma 2023), as well as a number of spatial **tidyverse** (Wickham et al. 2019; Wickham 2022) extensions, and a number of spatial analysis and visualisation packages that can be used with these packages, including **gstat** (Pebesma 2004; Pebesma and Graeler 2022), **spdep** (Bivand 2022c), **spatialreg** (Bivand and Piras 2022), **spatstat** (Baddeley, Rubak, and Turner 2015; Baddeley, Turner, and Rubak 2022), **tmap** (Tennekes 2018, 2022) and **mapview** (Appelhans et al. 2022).

Like data science, spatial data science seems to be a field that arises bottom-up in and from many existing scientific disciplines and industrial activities concerned with application of spatial data, rather than being a sub-discipline of an existing scientific discipline. Although there are various activities trying to scope it through focused conferences, symposia, chairs and study programs, we believe that the versatility of spatial data applications and questions will render such activity hard. Giving this book the title "spatial data science" is not another attempt to define the bounds of this field but rather an attempt to contribute to it from our 3-4 decades of experience working with researchers from various fields willing to publicly share research questions, data, and attempts to solve these questions with software. As a consequence, the selection of topics found in this book has a certain bias towards our own areas of research interest and experience. Platforms that have helped create an open research community include the ai-geostats and r-sig-geo mailing lists, sourceforge, r-forge, GitHub, and the OpenGeoHub summer schools organized yearly since 2006. The current possibility and willingness to cross data science language barriers opens a new and very exciting perspective. Our motivation to contribute to this field is a belief that open science leads to better science, and that better science might contribute to a more sustainable world.

Acknowledgements

We are grateful to the entire r-spatial community, especially those who

- developed r-spatial packages or contributed to their development
- contributed to discussions on twitter #rspatial or GitHub
- brought comments or asked questions in courses, summer schools, or conferences.

We are in particular grateful to Dewey Dunnington for implementing the **s2** package, and for active contributions from Sahil Bhandari, Jonathan Bahlmann for preparing the figures in Chapter 6, Claus Wilke, Jakub Nowosad, the "Spatial Data Science with R" classes of 2021 and 2022, and to those who actively contributed with GitHub issues, pull requests, or discussions:

- to the book repository (Nowosad, jonathom, JaFro96, singhkpratham, liuyadong, hurielreichel, PPaccioretti, Robinlovelace, Syverpet, jonas-hurst, angela-li, ALanguillaume, florisvdh, ismailsunni, andronaco),
- to the sf repository (aecoleman, agila5, andycraig, angela-li, ateucher, barry-rowlingson, bbest, BenGraeler, bhaskarvk, Bisaloo, bkmgit, christophertull, chrisyeh96, cmcaine, cpsievert, daissi, dankelley, DavisVaughan, dbaston, dblodgett-usgs, dcooley, demorenoc, dpprdan, drkrynstrng, etiennebr, famuvie, fdetsch, florisvdh, gregleleu, hadley, hughjonesd, huizezhang-sherry, jeffreyhanson, jeroen, jlacko, joethorley, joheisig, JoshOBrien, jwolfson, kadyb, karldw, kendonB, khondula, KHwong12, krlmlr, lambdamoses, lbusett, lcgodoy, lionel-, loicdtx, marwahaha, MatthieuStigler, mdsumner, MichaelChirico, microly, mpadge, mtennekes, nikolai-b, noerw, Nowosad, oliverbeagley, Pakillo, paleolimbot, pat-s, PPaccioretti, prdm0, ranghetti, rCarto, renejuan, rhijmans, rhurlin, rnuske, Robinlovelace, robitalec, rubak, rundel, statnmap, thomasp85, tim-salabim, tyluRp, uribo, Valexandre, wibeasley, wittja01, yutannihilation, Zedseayou),
- to the stars repository (a-benini, ailich, ateucher, btupper, dblodgett-usgs, djnavarro, ErickChacon, ethanwhite, etiennebr, flahn, floriandeboissieu, gavg712, gdkrmr, jannes-m, jeroen, JoshOBrien, kadyb, kendonB, mdsumner, michaeldorman, mtennekes, Nowosad, pat-s, PPaccioretti, przell, qdread, Rekyt, rhijmans, rubak, rushgeo, statnmap, uribo, yutannihilation),
- to the s2 repository (kylebutts, spiry34, jeroen, eddelbuettel).

Part I

Spatial Data

The first part of this book introduces concepts of spatial data science: maps, projections, vector and raster data structures, software, attributes and support, and data cubes. This part uses R only to generate text output or figures. The R code for this is not shown or explained, as it would distract from the message: Part II focuses on the use of R. The online version of this book, found at https://r-spatial.org/book/ contains the R code at the place where it is used in hidden sections that can be unfolded on demand and copied to the clipboard for execution and experimenting. Output from R code uses code font and has lines starting with a #, as in

```
# Linking to GEOS 3.11.1, GDAL 3.6.2, PROJ 9.1.1; sf_use_s2()
# is TRUE
```

More detailed explanation of R code to solve spatial data science problems starts in the second part of this book. Appendix B contains a short, elementary explanation of R data structures, Wickham (2014a) gives a more extensive treatment on this.

1

Getting Started

This chapter introduces a number of concepts associated with handling spatial and spatiotemporal data, pointing forward to later chapters where these concepts are discussed in more detail. It also introduces a number of open source technologies that form the foundation of all spatial data science language implementations.

1.1 A first map

The typical way to graph spatial data is by creating a map. Let us consider a simple map, shown in Figure 1.1.

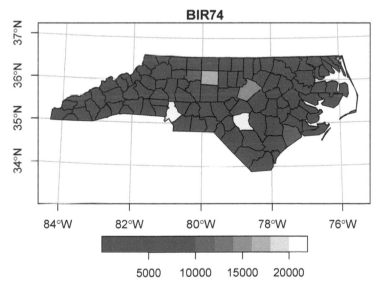

Figure 1.1: A first map: birth counts 1974-78, North Carolina counties

A number of graphical elements are present here, in this case:

- polygons are drawn with a black outline and filled with colours chosen according to a variable `BIR74`, whose name is in the title
- a legend key explains the meaning of the colours, and has a certain *colour palette* and *colour breaks*, values at which colour changes
- the background of the map shows curved lines with constant latitude or longitude (graticule)
- the axis ticks show the latitude and longitude values

Polygons are a particular form of *geometry*; spatial geometries (points, lines, polygons, pixels) are discussed in detail in Chapter 3. Polygons consist of sequences of points, connected by straight lines. How point locations of spatial data are expressed, or measured, is discussed in Chapter 2. As can be seen from Figure 1.1, lines of equal latitude and longitude do not form straight lines, indicating that some form of projection took place before plotting; projections are also discussed in Chapter 2 and Section 8.1.

The colour values in Figure 1.1 are derived from numeric values of a variable, `BIR74`, which has a single value associated with each geometry or *feature*. Chapter 5 discusses such feature attributes, and how they can relate to feature geometries. In this case, `BIR74` refers to birth counts, meaning counts *over the region*. This implies that the count does not refer to a value associated with every point inside the polygon, which the continuous colour might suggest, but rather measures an integral (sum) over the polygon.

Before plotting Figure 1.1 we had to read the data, in this case from a file (Section 7.1). Printing a data summary for the first three records of three attribute variables shows:

```
# Simple feature collection with 100 features and 3 fields
# Geometry type: MULTIPOLYGON
# Dimension:     XY
# Bounding box:  xmin: -84.3 ymin: 33.9 xmax: -75.5 ymax: 36.6
# Geodetic CRS:  NAD27
# # A tibble: 100 x 4
#    AREA BIR74 SID74                                        geom
#   <dbl> <dbl> <dbl>                          <MULTIPOLYGON [°]>
# 1 0.114  1091     1 (((-81.5 36.2, -81.5 36.3, -81.6 36.3, -8~
# 2 0.061   487     0 (((-81.2 36.4, -81.2 36.4, -81.3 36.4, -8~
# 3 0.143  3188     5 (((-80.5 36.2, -80.5 36.3, -80.5 36.3, -8~
# # ... with 97 more rows
```

The printed output shows:

- the (selected) dataset has 100 features (records) and 3 fields (attributes)
- the geometry type is `MULTIPOLYGON` (Chapter 3)

- it has dimension XY, indicating that each point will consist of 2 coordinate values
- the range of x and y values of the geometry
- the coordinate reference system (CRS) is geodetic, with coordinates in degrees longitude and latitude associated to the NAD27 datum (Chapter 2)
- the three selected attribute variables are followed by a variable geom of type MULTIPOLYGON with unit degrees that contains the polygon information

More complicated plots can involve facet plots with a map in each facet, as shown in Figure 1.2.

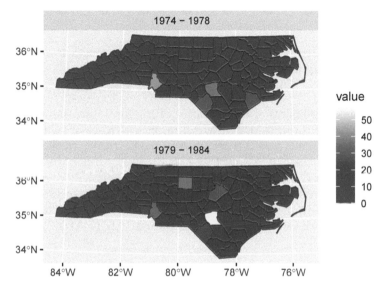

Figure 1.2: Facet maps of sudden infant death syndrome counts, 1974-78 and 1979-84, North Carolina counties

An interactive, leaflet-based map is obtained in Figure 1.3.

1.2 Coordinate reference systems

In Figure 1.1, the grey lines denote the *graticule*, a grid with lines along constant latitude or longitude. Clearly, these lines are not straight, which indicates that a *projection* of the data was used for which the x and y axis do not align with longitude and latitude. In Figure 1.3 we see that the north boundary of North Carolina is plotted as a straight line again, indicating that another projection was used.

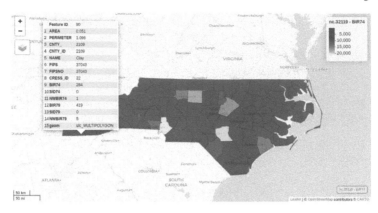

Figure 1.3: Interactive map created with **mapview**, showing feature attributes for a selected county in a popup window.

The ellipsoidal coordinates of the graticule of Figure 1.1 are associated with a particular *datum* (here: NAD27), which implicates a set of rules, what the shape of the Earth is and how it is attached to the Earth (to which point of the Earth is the origin associated, and how is it directed.) If one would measure coordinates with a GPS device (such as a mobile phone) it would typically report coordinates associated with the World Geodetic System 1984 (WGS84) datum, which can be around 30 m different from the identical coordinate values when associated with the North American Datum 1927 (NAD27).

Projections describe how we go back and forth between

- **ellipsoidal coordinates** which are expressed as degrees latitude and longitude, pointing to locations on a shape approximating the Earth's shape (ellipsoid or spheroid), and

- **projected coordinates** which are coordinates on a flat, two-dimensional coordinate system, used when plotting maps.

Datums transformations are associated with moving from one datum to another. Both topics are covered by *spatial reference systems*, and are described in more detail in Chapter 2.

1.3   Raster and vector data

Polygon, point, and line geometries are examples of *vector* data: point coordinates describe the "exact" locations that can be anywhere. Raster data on the other hand describe data where values are aligned on a *raster*, meaning on

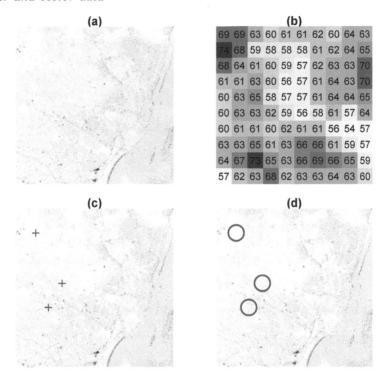

Figure 1.4: Raster maps (Olinda, Atlantic coast of Brazil): Landsat-7 blue band, with colour values derived from data values (a), the top-left 10×10 sub-image from (a) with numeric values shown (b), and overlayed by two different types of vector data: three sample points (c), and a 500 m radius around the points represented as polygons (d)

a regularly laid out lattice of usually square pixels. An example is shown in Figure 1.4.

Vector and raster data can be combined in different ways; for instance we can query the raster at the three points of Figure 1.4(c) or compute an aggregate, such as the average, over arbitrary regions such as the circles shown in Figure 1.4(d).

Other raster-to-vector conversions are discussed in Section 7.6 and include:

- converting raster pixels into point values
- converting raster pixels into small polygons, possibly merging polygons with identical values ("polygonize")
- generating lines or polygons that delineate continuous pixel areas with a certain value *range* ("contour")

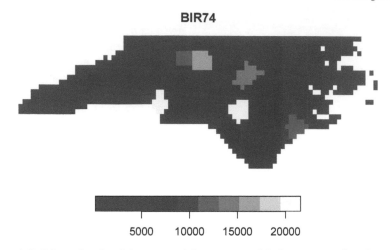

Figure 1.5: Map obtained by rasterizing county births counts for the period 1974-78 shown in 1.1

Vector-to-raster conversions can be as simple as rasterizing polygons, as shown in Figure 1.5. Other, more general vector-to-raster conversions that may involve statistical modelling include:

- interpolation of point values to points on a regular grid (Chapter 12)
- estimating densities of points over a regular grid (Chapter 11)
- area-weighted interpolation of polygon values to grid cells (Section 5.3)
- direct rasterization of points, lines, or polygons (Section 7.6)

1.4 Raster types

Raster dimensions describe how the rows and columns relate to spatial coordinates. Figure 1.6 shows a number of different possibilities.

Regular rasters like those shown in Figure 1.6 have a constant, not necessarily square cell size and axes aligned with the x and y (Easting and Northing) axes. Other raster types include those where the axes are no longer aligned with x and y (*rotated*), where axes are no longer perpendicular (*sheared*), or where cell size varies along a dimension (*rectilinear*). Finally, *curvilinear* rasters have cell size and/or direction properties that are no longer independent from the other raster dimension.

When a raster that is regular in a given coordinate reference system is projected to another raster while keeping each raster cell intact, it changes shape and may become rectilinear (for instance when going from ellipsoidal coordinates to Mercator, as in Figure 1.3) or curvilinear (for instance when going from

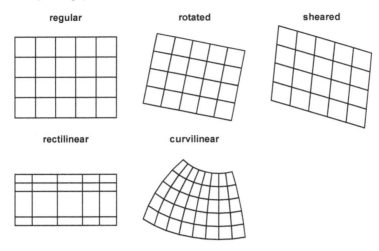

Figure 1.6: Various raster geometry types

ellipsoidal coordinates to Lambert Conic Conformal, as used in Figure 1.1). When reverting this procedure, one can recover the exact original raster.

Creating a new, regular grid in the new projection is called raster (or image) *reprojection* or *warping* (Section 7.8). Warping is lossy, irreversible, and needs to be informed whether raster cells should be interpolated, averaged or summed, or whether resampling using nearest neighbours should be used. For such choices it matters whether cell values reflect a categorical or continuous variable (see also Section 1.6).

1.5 Time series, arrays, data cubes

A lot of spatial data is not *only* spatial, but also temporal. Just like any observation is associated with an observation location, it is associated with an observation time or period. The dataset on the North Carolina counties shown above contains disease cases counted over two time periods (shown in Figure 1.2). Although the original dataset has these variables in two different columns, for plotting them these columns had to be stacked first, while repeating the associated geometries - a form called *tidy* Wickham (2014b). When we have longer time series associated with geometries, neither option - distributing time over multiple columns, or stacking columns while repeating geometries - works well, and a more effective way of storing such data would be a matrix or array, where one dimension refers to time, and the other(s) to space. The natural way for image or raster data is already to store them in matrices; time series of rasters then lead to a three-dimensional array. The general term for such data is a (spatiotemporal) **data cube**, where cube refers to arrays with any number of dimensions. Data cubes can refer to both raster and vector data, examples are given in Chapter 6.

1.6 Support

When we have spatial data with geometries that are not points but collections of points (multi-points, lines, polygons, pixels), then an attribute associated with these geometries has one of several different relationships to them. An attribute can have:

- a **constant** value for every point of the geometry
- a single value that is an **aggregate** over all points of the geometry
- a value that is unique to only this geometry, describing its **identity**

An example of a constant is land use or bedrock type of a polygon. An example of an aggregate is the number of births of a county over a given period of time. An example of an identity is a county name.

The spatial area an attribute value refers to is called its **support**: aggregate properties have "block" (or area, or line) support, constant properties have "point" support (they apply to every point). Support matters when we manipulate the data. For instance, Figure 1.5 was derived from a variable that has polygon support: the number of births per county. Rasterizing these values gives pixels with values that remain associated with counties. The result of the rasterization is a meaningless map: the numeric values ("birth totals") are not associated with the raster cells, and the county boundaries are no longer present. Totals of birth for the whole state or birth densities can no longer be recovered from the pixel values. Ignoring support can easily lead to meaningless results. Chapter 5 discusses this further.

Raster cell values may have point support or block support. An example of point support is elevation, when cells record the elevation of the point at the cell centre in a digital elevation model. An example of block (or cell) support is a satellite image pixel that gives the colour values averaged over (an area similar to) a pixel. Most file formats do not provide this information, yet it may be important to know when aggregating, regridding or warping rasters (Section 7.8), extracting values at point locations.

1.7   Spatial data science software

Although this book largely uses R and R packages for spatial data science, a number of these packages use software libraries that were not developed for R specifically. As an example, the dependency of R package **sf** on other R packages and system libraries is shown in Figure 1.7.

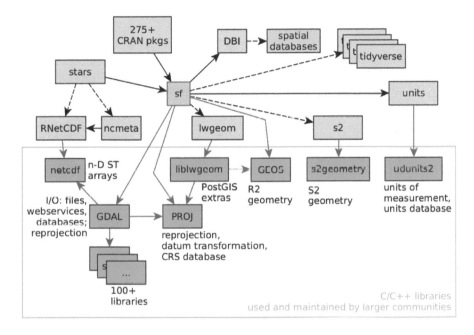

Figure 1.7: **sf** and its dependencies; arrows indicate strong dependency, dashed arrows weak dependency

The C or C++ libraries used (GDAL, GEOS, PROJ, liblwgeom, s2geometry, NetCDF, udunits2) are all developed, maintained, and used by (spatial) data science communities that are large and mostly different from the R community. By using these libraries, R users share how we understand what we are doing with these other communities. Because R, Python, and Julia provide interactive interfaces to this software, many users get closer to these libraries than do users of other software based on these libraries. The first part of this book describes many of the concepts implemented in these libraries, which is relevant to spatial data science in general.

1.7.1   GDAL

GDAL (Geospatial Data Abstraction Library) can be seen as the Swiss army knife of spatial data; besides for R it is being used in Python, QGIS, PostGIS, and more than 100 other software projects.

GDAL is a "library of libraries" – in order to read and write these data, it needs a large number of other libraries. It typically links to over 100 other libraries, each of which may provide access to a particular data file format, a database, a web service, or a particular compression codec.

Binary R packages distributed by CRAN contain only statically linked code: CRAN does not want to make any assumptions about presence of third-party libraries on the host system. As a consequence, when the `sf` package is installed in binary form from CRAN, it includes a copy of all the required external libraries as well as their dependencies, which may amount to 100 Mb.

1.7.2   PROJ

PROJ (or PRϕJ) is a library for cartographic projections and datum transformations: it converts spatial coordinates from one coordinate reference system to another. It comes with a large database of known projections and access to datum grids (high-precision, pre-calculated values for datum transformations). It aligns with an international standard for coordinate reference systems (Lott 2015). Chapter 2 deals with coordinate systems, and PROJ.

1.7.3   GEOS and s2geometry

GEOS (Geometry Engine Open Source) and s2geometry are two libraries for geometric operations. They are used to find measures (length, area, distance), and calculate predicates (do two geometries have any points in common?) or new geometries (which points do these two geometries have in common?). GEOS does this for flat, two-dimensional space (indicated by R^2), s2geometry does this for geometries on the sphere (indicated by S^2). Chapter 2 introduces coordinate reference systems, and Chapter 4 discusses more about the differences between working with these two spaces.

1.7.4   NetCDF, udunits2, liblwgeom

NetCDF (UCAR 2020) refers to a file format as well as a C library for reading and writing NetCDF files. It allows the definition of arrays of any dimensionality, and is widely used for spatial and spatiotemporal information, especially in the (climate) modelling communities. Udunits2 (UCAR 2014; Pebesma, Mailund, and Hiebert 2016; Pebesma et al. 2022) is a database and software library for units of measurement that allows the conversion of units, handles derived units, and supports user-defined units. The liblwgeom "library" is a software component of PostGIS (Obe and Hsu 2015) that contains several routines missing from GDAL or GEOS, including convenient access to GeographicLib routines (Karney 2013) that ship with PROJ.

1.8 Exercises

1. List five differences between raster and vector data.
2. In addition to those listed below Figure 1.1, list five further graphical components that are often found on a map.
3. In your own words, why is the numeric information shown in Figure 1.5 misleading (or meaningless)?
4. Under which conditions would you expect strong differences when doing geometrical operations on S^2, compared to doing them on R^2?

2

Coordinates

"Data are not just numbers, they are numbers with a context"; "In data analysis, context provides meaning" (Cobb and Moore 1997)

Before we can try to understand geometries like points, lines, polygons, coverage and grids, it is useful to review coordinate systems so that we have an idea what exactly coordinates of a point reflect. For spatial data, the location of observations are characterised by coordinates, and coordinates are defined in a coordinate system. Different coordinate systems can be used for this, and the most important difference is whether coordinates are defined over a 2 dimensional or 3 dimensional space referenced to orthogonal axes (Cartesian coordinates), or using distance and directions (polar coordinates, spherical and ellipsoidal coordinates). Besides a location of observation, all observations are associated with time of observation, and so time coordinate systems are also briefly discussed. First we will briefly review *quantities*, to learn what units and datum are.

2.1 Quantities, units, datum

The VIM ("International Vocabulary of Metrology", BIPM et al. (2012)) defines a *quantity* as a "property of a phenomenon, body, or substance, where the property has a magnitude that can be expressed as a number and a reference", where "[a] reference can be a measurement unit, a measurement procedure, a reference material, or a combination of such." Although one could argue about whether all data is constituted of quantities, there is no need to argue that proper data handling requires that numbers (or symbols) are accompanied by information on what they mean, in particular what they refer to.

A measurement system consists of *base units* for base quantities, and *derived units* for derived quantities. For instance, the SI system of units (Bureau International des Poids et Mesures 2006) consists of seven base units: length (metre, m), mass (kilogram, kg), time (second, s), electric current (ampere, A), thermodynamic temperature (Kelvin, K), amount of substance (mole, mol), and luminous intensity (candela, cd). Derived units are composed of products of integer powers of base units; examples are speed (m s^{-1}), density (kg m^{-3}) and area (m^2).

The special case of unitless measures can refer to either cases where units cancel out (for instance mass fraction: kg/kg, or angle measured in rad: m/m) or to cases where objects or events were counted (such as "5 apples"). Adding an angle to a count of apples would not make sense; adding 5 apples to 3 oranges may make sense if the result is reinterpreted in terms of a superclass, in this case as *pieces of fruit*. Many data variables have units that are not expressible as SI base units or derived units. Hand (2004) discusses many such measurement scales, including those used to measure variables like intelligence in social sciences, in the context of measurement units.

For many quantities, the natural origin of values is zero. This works for amounts, where differences between amounts result in meaningful negative values. For locations and times, differences have a natural zero interpretation: distance and duration. Absolute location (position) and time need a fixed origin, from which we can meaningfully measure other absolute space time points: we call this **a datum**. For space, a datum involves more than one dimension. The combination of a datum and a measurement unit (scale) is a *reference system*.

We will now elaborate how spatial locations can be expressed as either ellipsoidal or Cartesian coordinates. The next sections will deal with temporal and spatial reference systems, and how they are handled in R.

2.2   Ellipsoidal coordinates

Figure 2.1 shows both polar and Cartesian coordinates for a two-dimensional situation. In Cartesian coordinates, the point shown is $(x, y) = (3, 4)$, for polar

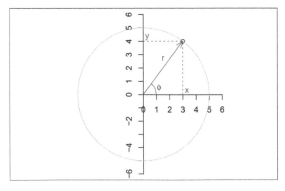

Figure 2.1: Two-dimensional polar (red) and Cartesian (blue) coordinates

coordinates it is $(r, \phi) = (5, \arctan(4/3))$, where $\arctan(4/3)$ is approximately 0.93 radians, or 53°. Note that x, y and r all have length units, where ϕ is an angle (a unitless length/length ratio). Converting back and forth between Cartesian and polar coordinates is trivial, as

$$x = r \cos\phi, \quad y = r \sin\phi, \text{ and}$$

$$r = \sqrt{x^2 + y^2}, \quad \phi = \text{atan2}(y, x)$$

where atan2 is used in favour of $\text{atan}(y/x)$ to take care of the right quadrant.

2.2.1 Spherical or ellipsoidal coordinates

In three dimensions, where Cartesian coordinates are expressed as (x, y, z), spherical coordinates are the three-dimensional equivalent of polar coordinates and can be expressed as (r, λ, ϕ), where:

- r is the radius of the sphere,
- λ is the longitude, measured in the (x, y) plane counter-clockwise from positive x, and
- ϕ is the latitude, the angle between the vector and the (x, y) plane.

Figure 2.2 illustrates Cartesian geocentric and ellipsoidal coordinates.

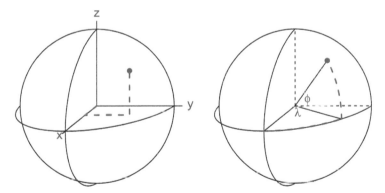

Figure 2.2: Cartesian geocentric coordinates (left) measure three distances, ellipsoidal coordinates (right) measure two angles, and possibly an ellipsoidal height

λ typically varies between $-180°$ and $180°$ (or alternatively from $0°$ to $360°$), ϕ from $-90°$ to $90°$. When we are only interested in points *on* a sphere with given radius, we can drop r: (λ, ϕ) now suffice to identify any point.

It should be noted that this is just *a* definition, one could for instance also choose to measure polar angle, the angle between the vector and z, instead of latitude. There is also a long tradition of specifying points as (ϕ, λ) but throughout this book we will stick to longitude-latitude, (λ, ϕ). The point denoted in Figure 2.2 has (λ, ϕ) or ellipsoidal coordinates with angular values

POINT (60 47)

measured in degrees, and geocentric coordinates

POINT Z (2178844 3773868 4641765)

measured in metres.

For points on an ellipse, there are two ways in which angle can be expressed (Figure 2.3): measured from the centre of the ellipse (ψ), or measured perpendicular to the tangent on the ellipse at the target point (ϕ).

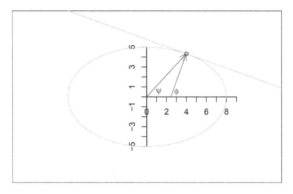

Figure 2.3: Angles on an ellipse: geodetic (blue) and geocentric (red) latitude

The most commonly used parametric model for the Earth is *an ellipsoid of revolution*, an ellipsoid with two equal semi-axes (Iliffe and Lott 2008). In effect, this is a flattened sphere (or spheroid): the distance between the poles is (slightly: about 0.33%) smaller than the distance between two opposite points on the equator. Under this model, longitude is always measured along a circle (as in Figure 2.2), and latitude along an ellipse (as in Figure 2.3). If we think of Figure 2.3 as a cross section of the Earth passing through the poles, the *geodetic* latitude measure ϕ is the one used when no further specification is given. The latitude measure ψ is called the *geocentric latitude*.

we can add *altitude* or elevation to longitude and latitude to define points that are above or below the ellipsoid, and obtain a three-dimensional space again. When defining altitude, we need to choose:

• where zero altitude is: on the ellipsoid, or relative to the surface approximating mean sea level (the geoid)?
• which direction is positive, and

- which direction is "straight up": perpendicular to the ellipsoid surface, or in the direction of gravity, perpendicular to the surface of the geoid?

All these choices may matter, depending on the application area and required measurement accuracies.

The shape of the Earth is not a perfect ellipsoid. As a consequence, several ellipsoids with different shape parameters and bound to the Earth in different ways are being used. Such ellipsoids are called *datums*, and are briefly discussed in Section 2.3, along with *coordinate reference systems*.

2.2.2 Projected coordinates, distances

Because paper maps and computer screens are much more abundant and practical than globes, when we look at spatial data we see it *projected*: drawn on a flat, two-dimensional surface. Computing the locations in a two-dimensional space means that we work with *projected* coordinates. Projecting ellipsoidal coordinates means that shapes, directions, areas, or even all three, are distorted (Iliffe and Lott 2008).

Distances between two points p_i and p_j in Cartesian coordinates are computed as Euclidean distances, in two dimensions by

$$d_{ij} = \sqrt{(x_i - x_j)^2 + (y_i - y_j)^2}$$

with $p_i = (x_i, y_i)$ and in three dimensions by

$$d_{ij} = \sqrt{(x_i - x_j)^2 + (y_i - y_j)^2 + (z_i - z_j)^2}$$

with $p_i = (x_i, y_i, z_i)$. These distances represent the length of a *straight* line between two points i and j.

For two points on a circle, the length of the arc of two points $c_1 = (r, \phi_i)$ and $c_2 = (r, \phi_2)$ is

$$s_{ij} = r \, |\phi_1 - \phi_2| = r \, \theta$$

with θ the angle between ϕ_1 and ϕ_2 in radians. For very small values of θ, we will have $s_{ij} \approx d_{ij}$, because a small arc segment is nearly straight.

For two points $p_1 = (\lambda_1, \phi_1)$ and $p_2 = (\lambda_2, \phi_2)$ on a sphere with radius r', the *great circle distance* is the arc length between p_1 and p_2 on the circle that passes through p_1 and p_2 and has the centre of the sphere as its centre, and is given by $s_{12} = r \, \theta_{12}$ with

$$\theta_{12} = \arccos(\sin \phi_1 \cdot \sin \phi_2 + \cos \phi_1 \cdot \cos \phi_2 \cdot \cos(|\lambda_1 - \lambda_2|))$$

the angle between p_1 and p_2, in radians.

Arc distances between two points on a spheroid are more complicated to compute; a good discussion on the topic and an explanation of the method implemented in GeographicLib (part of PROJ) is given in Karney (2013).

To show that these distance measures actually give different values, we computed them for the distance Berlin - Paris. Here, `gc_` refers to ellipsoidal and spherical great circle distances, `str_` refers to straight line, Euclidean distances between Cartesian geocentric coordinates associated on the WGS84 ellipse and sphere:

```
# Units: [km]
#  gc_ellipse str_ellipse   gc_sphere  str_sphere
#      879.70     879.00      877.46     876.77
```

2.2.3   Bounded and unbounded spaces

Two-dimensional and three-dimensional Euclidean spaces (R^2 and R^3) are unbounded. Every line in this space has infinite length, and areas or volumes have no natural upper limit. In contrast, spaces defined on a circle (S^1) or sphere (S^2) define a bounded set: there may be infinitely many points but the length and area of the circle, and the radius, area and volume of a sphere are bounded.

This may sound trivial but leads to some interesting challenges when handling spatial data. A polygon on R^2 has unambiguously an inside and an outside. On a sphere, S^2, any polygon divides the sphere in two parts, and which of these two is to be considered inside and which outside is ambiguous and needs to be defined by the traversal direction. Chapter 4 will further discuss consequences when working with geometries on S^2.

2.3   Coordinate reference systems

We follow Lott (2015) when defining the following concepts (italics indicate literal quoting):

- a **coordinate system** is a *set of mathematical rules for specifying how coordinates are to be assigned to points,*
- a **datum** is a *parameter or set of parameters that define the position of the origin, the scale, and the orientation of a coordinate system,*
- a **geodetic datum** is a *datum describing the relationship of a two- or three-dimensional coordinate system to the Earth,* and
- a **coordinate reference system** is a *coordinate system that is related to an object by a datum; for geodetic and vertical datums, the object will be the Earth.*

A readable text that further explains these concepts is Iliffe and Lott (2008).

The Earth does not follow a regular shape. The topography of the Earth is of course known to vary strongly, but also the surface formed by constant gravity at mean sea level, the geoid, is irregular. A commonly used model that is fit to the geoid is an ellipsoid of revolution, which is an ellipse with two identical minor axes. Fitting such an ellipsoid to the Earth gives a datum. However, fitting it to different areas, or based on different sets of reference points gives different fits, and hence different datums: a datum can for instance be fixed to a particular tectonic plate (like the European Terrestrial Reference System 1989 (ETRS89)), others can be globally fit (like WGS84). More local fits lead to smaller approximation errors.

The definitions above imply that coordinates in degrees longitude and latitude only have a meaning and can only be interpreted unambiguously as Earth coordinates, when the datum they are associated with is given.

Note that for projected data, the data that *were* projected are associated with a reference ellipsoid (datum). Going from one projection to another *without* changing datum is called *coordinate conversion*, and passes through the ellipsoidal coordinates associated with the datum involved. This process is lossless and invertible: the parameters and equations associated with a *conversion* are not empirical. Recomputing coordinates in a new datum is called *coordinate transformation*, and is approximate: because datums are a result of model fitting, transformations between datums are models too that have been fit; the equations involved are empirical, and multiple transformation paths, based on different model fits and associated with different accuracies, are possible.

Plate tectonics imply that within a global datum, fixed objects may have coordinates that change over time, and that transformations from one datum to another may be time-dependent. Earthquakes are a cause of more local and sudden changes in coordinates. Local datums may be fixed to tectonic plates (such as ETRS89), or may be dynamic.

2.4   PROJ and mapping accuracy

Very few living people active in open source geospatial software can remember the time before PROJ. PROJ (Evenden 1990) started in the 1970s as a Fortran project, and was released in 1985 as a C library for cartographic projections. It came with command line tools for direct and inverse projections, and could be linked to software to let it support (re)projection directly. Originally, datums were considered implicit, and no datum transformations were allowed.

In the early 2000s, PROJ was known as PROJ.4, after its never-changing major version number. Amongst others motivated by the rise of GPS, the need for datum transformations increased and PROJ.4 was extended with rudimentary datum support. PROJ definitions for coordinate reference systems would look like this:

```
+proj=utm +zone=33 +datum=WGS84 +units=m +no_defs
```

where *key=value* pairs are preceded by a + and separated by a space. This form came to be known as "PROJ.4 string", since the PROJ project stayed at version 4.x for several decades. Other datums would come with fields like:

```
+ellps=bessel +towgs84=565.4,50.3,465.6,-0.399,0.344,-1.877,4.072
```

indicating another ellipse, as well as the seven (or three) parameters for transforming from this ellipse to WGS84 (the "World Geodetic System 1984" global datum once popularised by GPS), effectively defining the datum in terms of a transformation to WGS84.

Along with PROJ.4 came a set of databases with known (registered) projections, from which the best known is the European Petroleum Survey Group (EPSG) registry. National mapping agencies would provide (and update over time) their best guesses of +towgs84= parameters for national coordinate reference systems, and distribute through the EPSG registry, which was part of PROJ distributions. For some transformations, *datum grids* were available and distributed as part of PROJ.4: such grids are raster maps that provide for every location pre-computed values for the shift in longitude and latitude, or elevation, for a particular datum transformation.

In PROJ.4, every coordinate transformation had to go through a conversion to and from WGS84; even reprojecting data associated with a datum different from WGS84 had to go through a transformation to and from WGS84. The associated errors of up to 100 m were acceptable for mapping purposes for not too small areas, but some applications need higher accuracy transformations. Examples include precision agriculture, planning flights of UAV's, or object tracking.

In 2018, after a successful "GDAL Coordinate System Barn Raising" initiative, a number of companies profiting from the open source geospatial software stack supported the development of a more modern, mature coordinate transformation system in PROJ. Over a few years, PROJ.4 evolved through versions 5, 6, 7, 8 and 9 and was hence renamed into PROJ (or PRϕJ).

The most notable changes include:

- although PROJ.4 strings can still be used to initialise certain coordinate reference systems, they are no longer sufficient to represent all of them; a new format, WKT-2 (described in next section) replaces it
- WGS84 as a hub datum is dropped: coordinate transformations no longer need to go through a particular datum
- multiple conversion or transformation paths (so-called pipelines) to go from CRS A to CRS B are possible, and can be reported along with the associated accuracy; PROJ will by default use the most accurate one but user control is possible

- transformation pipelines can chain an arbitrary number of elementary transformation operations, including swapping of axes and unit transformations
- datum grids, of which there are now *many* more, are no longer distributed with the library but are accessible from a content delivery network (CDN); PROJ allows enabling and disabling network access to these grids and only downloads the section(s) of the grid actually needed, storing it in a cache on the user's machine for future use
- coordinate transformations receive support for epochs, time-dependent transformations (and hence: four-dimensional coordinates, including the source and target time)
- the set of files with registered coordinate reference systems is handled in an SQLite database
- instead of always handling axis order (longitude, latitude), when the authority defines differently this is now obeyed (but see Section 2.5 and Section 7.7.6)

All these points sound like massive improvements, and accuracies of transformation can be below 1 metre. An interesting point is the last: Where we could safely assume for many decades that spatial data with ellipsoidal coordinates would have axis order (longitude, latitude), this is no longer the case. We will see in Section 7.7.6 how to deal with this.

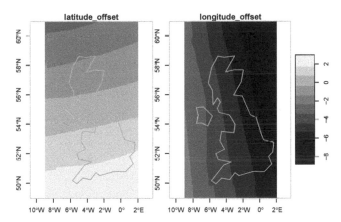

Figure 2.4: UK horizontal datum grid, from datum OSGB 1936 (EPSG:4277) to datum ETRS89 (EPSG:4258); units arc-seconds

Examples of a horizontal datum grids, downloaded from cdn.proj.org, are shown in Figure 2.4 and for a vertical datum grid in Figure 2.5. Datum grids may carry per-pixel accuracy values.

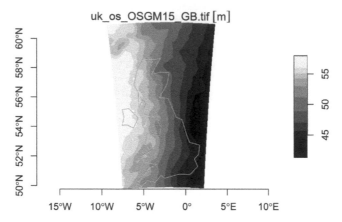

Figure 2.5: UK vertical datum grid, from ETRS89 (EPSG:4937) to ODN height (EPSG:5701), units m

2.5 WKT-2

Lott (2015) describes a standard for encoding coordinate reference systems, as well as transformations between them using *well-known text*; the standard (and format) is referred to informally as WKT-2. As mentioned above, GDAL and PROJ fully support this encoding. An example of WKT-2 for CRS `EPSG:4326` is:

```
GEOGCRS["WGS 84",
    ENSEMBLE["World Geodetic System 1984 ensemble",
        MEMBER["World Geodetic System 1984 (Transit)"],
        MEMBER["World Geodetic System 1984 (G730)"],
        MEMBER["World Geodetic System 1984 (G873)"],
        MEMBER["World Geodetic System 1984 (G1150)"],
        MEMBER["World Geodetic System 1984 (G1674)"],
        MEMBER["World Geodetic System 1984 (G1762)"],
        MEMBER["World Geodetic System 1984 (G2139)"],
        ELLIPSOID["WGS 84",6378137,298.257223563,
            LENGTHUNIT["metre",1]],
        ENSEMBLEACCURACY[2.0]],
    PRIMEM["Greenwich",0,
        ANGLEUNIT["degree",0.0174532925199433]],
    CS[ellipsoidal,2],
        AXIS["geodetic latitude (Lat)",north,
            ORDER[1],
            ANGLEUNIT["degree",0.0174532925199433]],
        AXIS["geodetic longitude (Lon)",east,
            ORDER[2],
            ANGLEUNIT["degree",0.0174532925199433]],
    USAGE[
```

```
      SCOPE["Horizontal component of 3D system."],
      AREA["World."],
      BBOX[-90,-180,90,180]],
   ID["EPSG",4326]]
```

This shows a coordinate system with the axis order *latitude, longitude,* although in most practical cases the axis order used is *longitude, latitude.* The *ensemble* of WGS84 ellipsoids listed represents its various updates over time. Ambiguity about *which* of these ensemble members a particular dataset should use leads to an uncertainty of several meters. The coordinate reference system OGC:CRS84 disambiguates the axis order and explicitly states it to be **longitude, latitude,** and is the recommended alternative to WGS84 datasets using this axis order. It does not disambiguate the datum ensemble problem.

A longer introduction on the history and recent changes in PROJ is given in Bivand (2020), building upon the work of Knudsen and Evers (2017) and Evers and Knudsen (2017).

2.6 Exercises

Try to solve the following exercises with R (without loading packages); try to use functions where appropriate:

1. list three *geographic* measures that do not have a natural zero origin
2. convert the (x, y) points $(10, 2)$, $(-10, -2)$, $(10, -2)$, and $(0, 10)$ to polar coordinates
3. convert the polar (r, ϕ) points $(10, 45°)$, $(0, 100°)$, and $(5, 359°)$ to Cartesian coordinates
4. assuming the Earth is a sphere with a radius of 6371 km, compute for (λ, ϕ) points the great circle distance between $(10, 10)$ and $(11, 10)$, between $(10, 80)$ and $(11, 80)$, between $(10, 10)$ and $(10, 11)$, and between $(10, 80)$ and $(10, 81)$ (units: degree). What are the distance units?

3

Geometries

Having learned how we represent coordinates systems, we can define how geometries can be described using these coordinate systems. This chapter will explain:

- *simple features*, a standard that describes point, line, and polygon geometries along with operations on them,
- operations on geometries,
- coverages, functions of space or space-time,
- tesselations, sub-divisions of larger regions into sub-regions, and
- networks.

Geometries on the sphere are discussed in Chapter 4, rasters and other rectangular sub-divisions of space or space time are discussed in Chapter 6.

3.1 Simple feature geometries

Simple feature geometries are a way to describe the geometries of *features*. By *features* we mean *things* that have a geometry, potentially implicitly some time properties, and further *attributes* that could include labels describing the thing and/or values quantitatively measuring it. The main application of simple feature geometries is to describe geometries in two-dimensional space by points, lines, or polygons. The "simple" adjective refers to the fact that the line or polygon geometries are represented by sequences of points connected with straight lines that do not self-intersect.

Simple features access is a standard (Herring 2011, 2010; ISO 2004) for describing simple feature geometries. It includes:

- a class hierarchy
- a set of operations
- binary and text encodings

We will first discuss the seven most common simple feature geometry types.

3.1.1 The big seven

The most common simple feature geometries used to represent a *single* feature are:

type	description
POINT	single point geometry
MULTIPOINT	set of points
LINESTRING	single linestring (two or more points connected by straight lines)
MULTILINESTRING	set of linestrings
POLYGON	exterior ring with zero or more inner rings, denoting holes
MULTIPOLYGON	set of polygons
GEOMETRYCOLLECTION	set of the geometries above

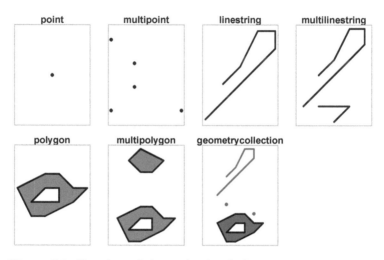

Figure 3.1: Sketches of the main simple feature geometry types

Figure 3.1 shows examples of these basic geometry types. The human-readable, "well-known text" (WKT) representation of the geometries plotted are:

```
POINT (0 1)
MULTIPOINT ((1 1), (2 2), (4 1), (2 3), (1 4))
LINESTRING (1 1, 5 5, 5 6, 4 6, 3 4, 2 3)
MULTILINESTRING ((1 1, 5 5, 5 6, 4 6, 3 4, 2 3), (3 0, 4 1, 2 1))
POLYGON ((2 1, 3 1, 5 2, 6 3, 5 3, 4 4, 3 4, 1 3, 2 1),
    (2 2, 3 3, 4 3, 4 2, 2 2))
MULTIPOLYGON (((2 1, 3 1, 5 2, 6 3, 5 3, 4 4, 3 4, 1 3, 2 1),
    (2 2, 3 3, 4 3, 4 2, 2 2)), ((3 7, 4 7, 5 8, 3 9, 2 8, 3 7)))
GEOMETRYCOLLECTION (
    POLYGON ((2 1, 3 1, 5 2, 6 3, 5 3, 4 4, 3 4, 1 3, 2 1),
    (2 2 , 3 3, 4 3, 4 2, 2 2)),
    LINESTRING (1 6, 5 10, 5 11, 4 11, 3 9, 2 8),
    POINT (2 5),
    POINT (5 4)
)
```

In this representation, coordinates are separated by space, and points by commas. Sets are grouped by parentheses, and separated by commas. Polygons consist of an outer ring followed by zero or more inner rings denoting holes.

Individual points in a geometry contain at least two coordinates: x and y, in that order. If these coordinates refer to ellipsoidal coordinates, x and y usually refer to longitude and latitude, respectively, although sometimes to latitude and longitude (see Section 2.4 and Section 7.7.6).

3.1.2 Simple and valid geometries, ring direction

Linestrings are called *simple* when they do not self-intersect:

```
# LINESTRING (0 0, 1 1, 2 2, 0 2, 1 1, 2 0)

# is_simple
#     FALSE
```

Valid polygons and multi-polygons obey all of the following properties:

- polygon rings are closed (the last point equals the first)
- polygon holes (inner rings) are inside their exterior ring
- polygon inner rings maximally touch the exterior ring in single points, not over a line
- a polygon ring does not repeat its own path
- in a multi-polygon, an external ring maximally touches another exterior ring in single points, not over a line

If this is not the case, the geometry concerned is not valid. Invalid geometries typically cause errors when they are processed, but they can usually be repaired to make them valid.

A further convention is that the outer ring of a polygon is winded counter-clockwise, while the holes are winded clockwise, but polygons for which this is not the case are still considered valid. For polygons on the sphere, the "clockwise" concept is not very useful: if for instance we take the equator as polygon, is the Northern Hemisphere or the Southern Hemisphere "inside"? The convention taken here is to consider the area on the left while traversing the polygon is considered the polygon's inside (see also Section 7.3).

3.1.3 Z and M coordinates

In addition to X and Y coordinates, single points (vertices) of simple feature geometries may have:

- a Z coordinate, denoting altitude, and/or
- an M value, denoting some "measure"

The M attribute shall be a property of the vertex. It sounds attractive to encode a time stamp in it for instance to pack movement data (trajectories) in LINESTRINGs.

These become however invalid (or "non-simple") once the trajectory self-intersects, which happens when only X and Y are considered for self-intersections.

Both Z and M are not found often, and software support to do something useful with them is (still) rare. Their WKT representations are fairly easily understood:

```
# POINT Z (1 3 2)
```

```
# POINT M (1 3 2)
```

```
# LINESTRING ZM (3 1 2 4, 4 4 2 2)
```

3.1.4 Empty geometries

A very important concept in the feature geometry framework is that of the empty geometry. Empty geometries arise naturally when we do geometrical operations (Section 3.2), for instance when we want to know the intersection of POINT (0 0) and POINT (1 1):

```
# GEOMETRYCOLLECTION EMPTY
```

and it represents essentially the empty set: when combining (unioning) an empty point with other non-empty geometries, it vanishes.

All geometry types have a special value representing the empty (typed) geometry, like

```
# POINT EMPTY
```

```
# LINESTRING M EMPTY
```

and so on, but they all point to the empty set, differing only in their dimension (Section 3.2.2).

3.1.5 Ten further geometry types

There are 10 more geometry types which are more rare, but increasingly find implementation:

type	description
CIRCULARSTRING	The CircularString is the basic curve type, similar to a LineString in the linear world. A single segment requires three points, the start and end points (first and third) and any other point on the arc. The exception to this is for a closed circle, where the start and end points are the same. In this case the second point MUST be the centre of the arc, i.e., the opposite side of the circle. To chain arcs together, the last point of the previous arc becomes the first point of the next arc, just like in LineString. This means that a valid circular string must have an odd number of points greater than 1.

type	description
COMPOUNDCURVE	A CompoundCurve is a single, continuous curve that has both curved (circular) segments and linear segments. That means that in addition to having well-formed components, the end point of every component (except the last) must be coincident with the start point of the following component.
CURVEPOLYGON	Example compound curve in a curve polygon: `CURVEPOLYGON( COMPOUNDCURVE( CIRCULARSTRING(0 0,2 0, 2 1, 2 3, 4 3),(4 3, 4 5, 1 4, 0 0)), CIRCULARSTRING(1.7 1, 1.4 0.4, 1.6 0.4, 1.6 0.5, 1.7 1))`
MULTICURVE	A MultiCurve is a 1 dimensional GeometryCollection whose elements are Curves. It can include linear strings, circular strings, or compound strings.
MULTISURFACE	A MultiSurface is a 2 dimensional GeometryCollection whose elements are Surfaces, all using coordinates from the same coordinate reference system.
CURVE	A Curve is a 1 dimensional geometric object usually stored as a sequence of Points, with the subtype of Curve specifying the form of the interpolation between Points
SURFACE	A Surface is a 2 dimensional geometric object
POLYHEDRALSURFACE	A PolyhedralSurface is a contiguous collection of polygons, which share common boundary segments
TIN	A TIN (triangulated irregular network) is a PolyhedralSurface consisting only of Triangle patches.
TRIANGLE	A Triangle is a polygon with three distinct, non-collinear vertices and no interior boundary

CIRCULARSTRING, COMPOUNDCURVE and CURVEPOLYGON are not described in the SFA standard, but in the SQL-MM part 3 standard. The descriptions above were copied from the PostGIS manual.

3.1.6 Text and binary encodings

Part of the simple feature standard are two encodings: a text and a binary encoding. The well-known text encoding, used above, is human-readable. The well-known binary encoding is machine-readable. Well-known binary (WKB) encodings are lossless and typically faster to work with than text encoding (and decoding), and they are used for instance in all communications between R package **sf** and the GDAL, GEOS, liblwgeom, and s2geometry libraries (Figure 1.7).

3.2   Operations on geometries

Simple feature geometries can be queried for properties, or transformed or combined
into new geometries, and combinations of geometries can be queried for further
properties. This section gives an overview of the operations entirely focusing on
geometrical properties. Chapter 5 focuses on the analysis of non-geometrical feature
properties, in relationship to their geometries. Some of the material in this section
appeared in Pebesma (2018).

We can categorise operations on geometries in terms of what they take as input, and
what they return as output. In terms of output we have operations that return:

- **predicates**: a logical asserting a certain property is `TRUE`
- **measures**: a quantity (a numeric value, possibly with measurement unit)
- **transformations**: newly generated geometries

and in terms of what they operate on, we distinguish operations that are:

- **unary** when they work on a single geometry
- **binary** when they work on pairs of geometries
- **n-ary** when they work on sets of geometries

3.2.1   Unary predicates

Unary predicates describe a certain property of a geometry. The predicates `is_simple`,
`is_valid`, and `is_empty` return respectively whether a geometry is simple, valid,
or empty. Given a coordinate reference system, `is_longlat` returns whether the
coordinates are geographic or projected. `is(geometry, class)` checks whether a
geometry belongs to a particular class.

3.2.2   Binary predicates and DE-9IM

The Dimensionally Extended Nine-Intersection Model (DE-9IM, Clementini, Di
Felice, and Oosterom 1993; Egenhofer and Franzosa 1991) is a model that describes
the qualitative relation between any two geometries in two-dimensional space (R^2).
Any geometry has a *dimension* value that is:

- 0 for points,
- 1 for linear geometries,
- 2 for polygonal geometries, and
- F (false) for empty geometries

Any geometry also has an inside (I), a boundary (B), and an exterior (E); these roles
are obvious for polygons, however, for:

- **lines** the boundary is formed by the end points, and the interior by all non-end
  points on the line

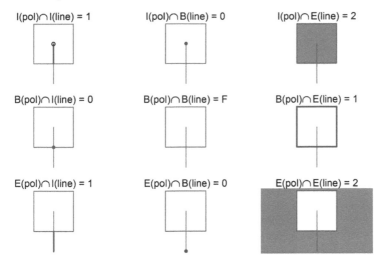

Figure 3.2: DE-9IM: intersections between the interior, boundary, and exterior of a polygon (rows) and of a linestring (columns) indicated by red

- **points** have a zero-dimensional inside but no boundary

Figure 3.2 shows the intersections between the I, B, and E components of a polygon and a linestring indicated by red; the sub-plot title gives the dimension of these intersections (0, 1, 2 or F). The relationship between the polygon and the line geometry is the concatenation of these dimensions:

```
#       [,1]
# [1,]  "1020F1102"
```

where the first three characters are associated with the inside of the *first* geometry (polygon): Figure 3.2 is summarised row-wise. Using this ability to express relationships, we can also query pairs of geometries about particular conditions expressed in a *mask string*. As an example, the string "*0*******" would evaluate TRUE when the second geometry has one or more boundary *points* in common with the interior of the first geometry; the symbol * standing for "any dimensionality" (0, 1, 2 or F). The mask string "T********" matches pairs of geometry with intersecting interiors, where the symbol T stands for any non-empty intersection of dimensionality 0, 1, or 2.

Binary predicates are further described using normal-language verbs, using DE-9IM definitions. For instance, the predicate equals corresponds to the relationship "T*F**FFF*". If any two geometries obey this relationship, they are (topologically) equal, but may have a different ordering of nodes.

A list of binary predicates is:

predicate	meaning	inverse of
`contains`	None of the points of A are outside B	`within`
`contains_properly`	A contains B and B has no points in common with the boundary of A	
`covers`	No points of B lie in the exterior of A	`covered_by`
`covered_by`	Inverse of `covers`	
`crosses`	A and B have some but not all interior points in common	
`disjoint`	A and B have no points in common	`intersects`
`equals`	A and B are topologically equal: node order or number of nodes may differ; identical to A contains B and A within B	
`equals_exact`	A and B are geometrically equal, and have identical node order	
`intersects`	A and B are not disjoint	`disjoint`
`is_within_distance`	A is closer to B than a given distance	
`within`	None of the points of B are outside A	`contains`
`touches`	A and B have at least one boundary point in common, but no interior points	
`overlaps`	A and B have some points in common; the dimension of these is identical to that of A and B	
`relate`	Given a mask pattern, return whether A and B adhere to this pattern	

The Wikipedia DE-9IM page provides the `relate` patterns for each of these verbs. They are important to check out; for instance *covers* and *contains* (and their inverses) are often not completely intuitive:

- if A *contains* B, B has no points in common with the exterior *or boundary* of A
- if A *covers* B, B has no points in common with the exterior of A

3.2.3   Unary measures

Unary measures return a measure or quantity that describes a property of the geometry:

measure	returns
`dimension`	0 for points, 1 for linear, 2 for polygons, possibly `NA` for empty geometries
`area`	the area of a geometry
`length`	the length of a linear geometry

3.2.4 Binary measures

distance returns the distance between pairs of geometries. The qualitative measure **relate** (without mask) gives the relation pattern. A description of the geometrical relationship between two geometries is given in Section 3.2.2.

3.2.5 Unary transformers

Unary transformations work on a per-geometry basis, and return for each geometry a new geometry.

transformer	returns a geometry …
centroid	of type **POINT** with the geometry's centroid
buffer	that is larger (or smaller) than the input geometry, depending on the buffer size
jitter	that was moved in space a certain amount, using a bivariate uniform distribution
wrap_dateline	cut into pieces that no longer cover or cross the dateline
boundary	with the boundary of the input geometry
convex_hull	that forms the convex hull of the input geometry (Figure 3.3)
line_merge	after merging connecting **LINESTRING** elements of a **MULTILINESTRING** into longer **LINESTRING**s.
make_valid	that is valid
node	with added nodes to linear geometries at intersections without a node; only works on individual linear geometries
point_on_surface	with a (arbitrary) point on a surface
polygonize	of type polygon, created from lines that form a closed ring
segmentize	a (linear) geometry with nodes at a given density or minimal distance
simplify	simplified by removing vertices/nodes (lines or polygons)
split	that has been split with a splitting linestring
transform	transformed or convert to a new coordinate reference system (Chapter 2)
triangulate	with Delauney triangulated polygon(s) (Figure 3.3)
voronoi	with the Voronoi tessellation of an input geometry (Figure 3.3)
zm	with removed or added **Z** and/or **M** coordinates
collection_extract	with sub-geometries from a **GEOMETRYCOLLECTION** of a particular type
cast	that is converted to another type
+	that is shifted over a given vector
*	that is multiplied by a scalar or matrix

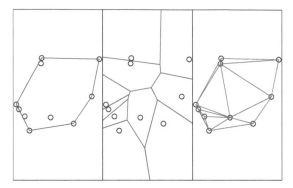

Figure 3.3: For a set of points, left: convex hull (red); middle: Voronoi polygons; right: Delauney triangulation

3.2.6 Binary transformers

Binary transformers are functions that return a geometry based on operating on a pair of geometries. They include:

function	returns	infix operator
intersection	the overlapping geometries for pair of geometries	&
union	the combination of the geometries; removes internal boundaries and duplicate points, nodes or line pieces	|
difference	the geometries of the first after removing the overlap with the second geometry	/
sym_difference	the combinations of the geometries after removing where they intersect; the negation (opposite) of intersection	%/%

3.2.7 N-ary transformers

N-ary transformers operate on sets of geometries. union can be applied to a set of geometries to return its geometrical union. Otherwise, any set of geometries can be combined into a MULTI-type geometry when they have equal dimension, or else into a GEOMETRYCOLLECTION. Without unioning, this may lead to a geometry that is not valid, for instance when two polygon rings have a boundary line in common.

N-ary intersection and difference take a single argument but operate (sequentially) on all pairs, triples, quadruples, etc. Consider the plot in Figure 3.4: how do we identify the area where all three boxes overlap? Using binary intersections gives us intersections for all pairs: 1-1, 1-2, 1-3, 2-1, 2-2, 2-3, 3-1, 3-2, 3-3, but that does not let us identify areas where more than two geometries intersect. Figure 3.4 (right) shows the n-ary intersection: the seven unique, non-overlapping geometries originating from intersection of one, two, *or more* geometries.

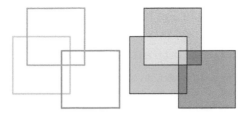

Figure 3.4: Left: three overlapping squares – how do we identify the small box where all three overlap? Right: unique, non-overlapping n-ary intersections

Similarly, one can compute an n-ary *difference* from a set $\{s_1, s_2, s_3, ...\}$ by creating differences $\{s_1, s_2 - s_1, s_3 - s_2 - s_1, ...\}$. This is shown in Figure 3.5, (left) for the original set, and (right) for the set after reversing its order to make clear that the result here depends on the ordering of the input geometries. Again, resulting geometries do not overlap.

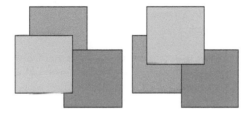

Figure 3.5: Difference between subsequent boxes, left: in original order; right: in reverse order

3.3 Precision

Geometrical operations, such as finding out whether a certain point is on a line, may fail when coordinates are represented by double precision floating point numbers, such as 8-byte doubles used in R. An often chosen remedy is to limit the precision of the coordinates before the operation. For this, a *precision model* is adopted; the most common is to choose a factor p and compute rounded coordinates c' from original coordinates c by

$$c' = \text{round}(p \cdot c)/p$$

Rounding of this kind brings the coordinates to points on a regular grid with spacing $1/p$, which is beneficial for geometric computations. Of course, it also affects all computations like areas and distances, and may turn valid geometries into invalid ones. Which precision values are best for which application is often a matter of common sense combined with trial and error.

3.4   Coverages: tessellations and rasters

The Open Geospatial Consortium defines a *coverage* as a "feature that acts as a function to return values from its range for any direct position within its spatiotemporal domain" (Baumann, Hirschorn, and Masó 2017). Having a *function* implies that for every space time "point", every combination of a spatial point and a moment in time of the spatiotemporal domain, we have a *single* value for the range. This is a very common situation for spatiotemporal phenomena, a few examples can be given:

- boundary disputes aside, at a give time every point in a region (domain) belongs to a single administrative unit (range)
- at any given moment in time, every point in a region (domain) has a certain *land cover type* (range)
- every point in an area (domain) has a single surface elevation (range), which could be measured with respect to a given mean sea level surface
- every spatiotemporal point in a three-dimensional body of air (domain) has single value for temperature (range)

A caveat here is that because observation or measurement always takes time and requires space, measured values are always an average over a spatiotemporal volume, and hence range variables can rarely be measured for true, zero-volume "points"; for many practical cases however the measured volume is small enough to be considered a "point". For a variable like *land cover type* the volume needs to be chosen such that the types distinguished make sense with respect to the measured areal units.

In the first two of the given examples the range variable is *categorical*, in the last two the range variable is *continuous*. For categorical range variables, if large connected areas have a constant range value, an efficient way to represent these data is by storing the boundaries of the areas with constant value, such as country boundaries. Although this can be done (and is often done) by a set of simple feature geometries (polygons or multi-polygons), this brings along some challenges:

- it is hard to guarantee for such a set of simple feature polygons that they do not overlap, or that there are no unwanted gaps between them
- simple features have no way of assigning points *on* the boundary of two adjacent polygons uniquely to a single polygon, which conflicts with the interpretation as coverage

3.4.1   Topological models

A data model that guarantees no inadvertent gaps or overlaps of polygonal coverages is the *topological* model, examples of which are found in geographic information systems (GIS) like GRASS GIS or ArcGIS. Topological models store boundaries between polygons only once and register which polygonal area is on either side of a boundary.

Deriving the set of (multi)polygons for each area with a constant range value from a topological model is straightforward; the other way around, reconstructing topology from a set of polygons typically involves setting thresholds on errors and handling gaps or overlaps.

3.4.2 Raster tessellations

A tessellation is a sub-division of a space (area, volume) into smaller elements by ways of polygons. A regular tessellation does this with regular polygons: triangles, squares, or hexagons. Tessellations using squares are commonly used for spatial data and are called *raster data*. Raster data tessellate each spatial dimension d into regular cells, formed by left-closed and right-open intervals d_i:

$$d_i = d_0 + [i \times \delta, (i+1) \times \delta) \tag{3.1}$$

with d_0 an offset, δ the interval (cell or pixel) size, and where the cell index i is an arbitrary but consecutive set of integers. The δ value is often taken negative for the y-axis (Northing), indicating that raster row numbers increasing Southwards correspond to y-coordinates increasing Northwards.

Whereas in arbitrary polygon tessellations the assignment of points to polygons is ambiguous for points falling on a boundary shared by two polygons, using left-closed "[" and right-open ")" intervals in regular tessellations removes this ambiguity. This means that for rasters with negative δ values for the y coordinate and positive for the x-coordinate, only the top-left corner point is part of each raster cell. An artifact resulting from this is shown in Figure 3.6.

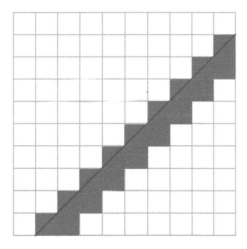

Figure 3.6: Rasterization artifact: as only top-left corners are part of the raster cell, only cells touching the red line below the diagonal line are rasterized

Tessellating the time dimension with left-closed right-open intervals is very common, and it reflects the implicit assumption underlying time series software such as the **xts** package in R, where time stamps indicate the start of time intervals. Different models can be combined: one could use simple feature polygons to tessellate space and combine this with a regular tessellation of time in order to cover a space time *vector data cube*. Raster and vector data cubes are discussed in Chapter 6.

As mentioned above, besides square cells the other two shapes that can lead to regular tessellations of R^2 are triangles and hexagons. On the sphere, there are a few more, including cube, octahedron, icosahedron, and dodecahedron. A spatial index that builds on the cube is s2geometry, the H3 library uses the icosahedron and densifies that with (mostly) hexagons. Mosaics that cover the entire Earth are also called *discrete global grids*.

3.5 Networks

Spatial networks are typically composed of linear (`LINESTRING`) elements, but possess further topological properties describing the network coherence:

- start- and end-points of a linestring may be connected to other linestring start or end points, forming a set of nodes and edges
- edges may be directed, to only allow for connection (flow, transport) in one way

R packages including **osmar** (Schlesinger and Eugster 2013), **stplanr** (Lovelace, Ellison, and Morgan 2022), and **sfnetworks** (van der Meer et al. 2022) provide functionality for constructing network objects, and working with them, including computation of shortest or fastest routes through a network. Package **spatstat** (Baddeley, Turner, and Rubak 2022; Baddeley, Rubak, and Turner 2015) has infrastructure for analysing point patterns on linear networks (Chapter 11). Chapter 12 of Lovelace, Nowosad, and Muenchow (2019) has a transportation application using networks.

3.6 Exercises

For the following exercises, use R where possible.

1. Give two examples of geometries in 2-D (flat) space that cannot be represented as simple feature geometries, and create a plot of them.
2. Recompute the coordinates 10.542, 0.01, 45321.6789 using precision values 1, 1e3, 1e6, and 1e-2.
3. Describe a practical problem for which an n-ary intersection would be needed.
4. How can you create a Voronoi diagram (Figure 3.3) that has one closed polygons for every single point?
5. Give the unary measure `dimension` for geometries `POINT Z (0 1 1)`, `LINESTRING Z (0 0 1,1 1 2)`, and `POLYGON Z ((0 0 0,1 0 0,1 1 0,0 0 0))`
6. Give the DE-9IM relation between `LINESTRING(0 0,1 0)` and `LINESTRING(0.5 0,0.5 1)`; explain the individual characters.
7. Can a set of simple feature polygons form a coverage? If so, under which constraints?

8. For the **nc** counties in the dataset that comes with R package **sf**, find the points touched by four counties.
9. How would Figure 3.6 look like if δ for the y-coordinate was positive?

4

Spherical Geometries

"*There are too many false conclusions drawn and stupid measurements made when geographic software, built for projected Cartesian coordinates in a local setting, is applied at the global scale*" (Chrisman 2012)

The previous chapter discussed geometries defined on the plane, R^2. This chapter discusses what changes when we consider geometries not on the plane, but on the sphere (S^2).

Although we learned in Chapter 2 that the shape of the Earth is usually approximated by an ellipsoid, none of the libraries shown in green in Figure 1.7 provide access to a comprehensive set of functions that compute on an ellipsoid. Only the s2geometry (Dunnington, Pebesma, and Rubak 2023; Veach et al. 2020) library does provide it using a sphere rather than an ellipsoid. However, when compared to using a flat (projected) space as we did in the previous chapter, a sphere is a *much* better approximation to an ellipsoid.

4.1 Straight lines

The basic premise of *simple features* of Chapter 3 is that geometries are represented by sequences of points *connected by straight lines*. On R^2 (or any Cartesian space), this is trivial, but on a sphere straight lines do not exist. The shortest line connecting two points is an arc of the circle through both points and the centre of the sphere, also called a *great circle segment*. A consequence is that "the" shortest distance line connecting two points on opposing sides of the sphere does not exist, as any great circle segment connecting them has equal length. Note that the GeoJSON standard (Butler et al. 2016) has its own definition of straight lines in geodetic coordinates (see Exercise 1 at the end of this chapter).

4.2 Ring direction and full polygon

Any polygon on the sphere divides the sphere surface in two parts with finite area: the inside and the outside. Using the "counter-clockwise rule" as was done for R^2 will not work because the direction interpretation depends on what is defined as inside.

A convention here is to define the inside as the left (or right) side of the polygon boundary when traversing its points in sequence. Reversal of the node order then switches inside and outside.

In addition to empty polygons, one can define the *full polygon* on a sphere, which comprises its entire surface. This is useful, for instance for computing the oceans as the geometric difference between the full polygon and the union of the land mass (see Figure 8.1 and Figure 11.6).

4.3   Bounding box, rectangle, and cap

Where in R^2 one can easily define bounding boxes as the range of the x and y coordinates, for ellipsoidal coordinates these ranges are not of much use when geometries cross the antimeridian (longitude +/- 180) or one of the poles. The assumption in R^2 that lower x values are Westwards of higher ones does not hold when crossing the antimeridian. An alternative to delineating an area on a sphere that is more natural is the *bounding cap*, defined by its centre coordinates and a radius. For Antarctica, as depicted in Figures -Figure 4.1 (a) and (c), the bounding box formed by coordinate ranges is

```
#    xmin    ymin    xmax    ymax
# -180.0   -85.2   179.6   -60.5
```

which clearly does not contain the region (`ymin` being -90 and `xmax` 180). Two geometries that do contain the region are the bounding cap:

```
#    lng lat angle
# 1    0 -90  29.5
```

and the bounding *rectangle*:

```
#    lng_lo lat_lo lng_hi lat_hi
# 1    -180    -90    180  -60.5
```

For an area spanning the antimeridian, here the Fiji island country, the bounding box:

```
#    xmin    ymin    xmax    ymax
# -179.9   -21.7   180.2   -12.5
```

seems to span most of the Earth, as opposed to the bounding rectangle:

```
#    lng_lo lat_lo lng_hi lat_hi
# 1     175  -21.7   -178  -12.5
```

where a value `lng_lo` *larger* than `lng_hi` indicates that the bounding rectangle spans the antimeridian. This property could not be inferred from the coordinate ranges.

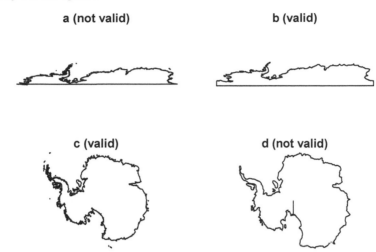

Figure 4.1: Antarctica polygon, (a, c): *not* passing through POINT(-180 -90);
(b, d): passing through POINT(-180 -90) and POINT(180 -90)

4.4   Validity on the sphere

Many global datasets are given in ellipsoidal coordinates but are prepared in a way
that they "work" when interpreted on the R^2 space [-180,180] × [-90,90]. This means
that:

- geometries crossing the antimeridian (longitude +/- 180) are cut in half, such that
  they no longer cross it (but nearly touch each other)
- geometries including a pole, like Antarctica, are cut at +/- 180 and make an
  excursion through -180,-90 and 180,-90 (both representing the Geographic South
  Pole)

Figure 4.1 shows two different representations of Antarctica, plotted with ellipsoidal
coordinates taken as R^2 (top) and in a Polar Stereographic projection (bottom),
without (left) and with (right) an excursion through the Geographic South Pole. In
the projections as plotted, polygons (b) and (c) are valid, polygon (a) is not valid as
it self-intersects, and polygon (d) is not valid because it traverses the same edge to
the South Pole twice. On the sphere (S^2), polygon (a) is valid but (b) is not, for the
same reason as (d) is not valid.

4.5   Exercises

For the following exercises, use R where possible or relevant.

1. How does the GeoJSON format (Butler et al. 2016) define "straight" lines between ellipsoidal coordinates (Section 3.1.1)? Using this definition of straight, how does **LINESTRING(0 85,180 85)** look like in an Arctic polar projection? How could this geometry be modified to have it cross the North Pole?
2. For a typical polygon on S^2, how can you find out ring direction?
3. Are there advantages of using bounding caps over using bounding boxes? If so, list them.
4. Why is, for small areas, the orthographic projection centred at the area a good approximation of the geometry as handled on S^2?
5. For   **rnaturalearth::ne_countries(country = "Fiji",** **returnclass = "sf")**, check whether the geometry is valid on R^2, on an orthographic projection centred on the country, and on S^2. How can the geometry be made valid on S^2? Plot the resulting geometry back on R^2. Compare the centroid of the country, as computed on R^2 and on S^2, and the distance between the two.
6. Consider dataset **gisco_countries** in R package **giscoR**, and select the country with **NAME_ENGL == "Fiji"**. Does it have a valid geometry on the sphere? If so, how was this accomplished?

5

Attributes and Support

Feature *attributes* refer to the properties of features ("things") that do not describe the feature's geometry. Feature attributes can be *derived* from geometry (such as length of a `LINESTRING` or area of a `POLYGON`), but they can also refer to non-derived properties, such as:

- name of a street or county
- number of people living in a country
- type of a road
- soil type in a polygon from a soil map
- opening hours of a shop
- body weight or heart beat rate of an animal
- NO_2 concentration measured at an air quality monitoring station

In some cases, time properties can be seen as attributes of features, for instance the date of birth of a person or the construction year of a road. When an attribute such as for instance air quality is a function of both space and time, time is best handled on equal footing with geometry (often in a data cube, see Chapter 6).

Spatial data science software implementing simple features typically organises data in tables that contain both geometries and attributes for features; this is true for **geopandas** in Python, **PostGIS** tables in PostgreSQL, and **sf** objects in R. The geometric operations described in Section 3.2 operate on geometries *only*, and may occasionally yield new attributes (predicates, measures, or transformations), but they do not operate on attributes present.

When, while manipulating geometries, attribute *values* are retained unmodified, support problems may arise. If we look into a simple case of replacing a county polygon with the centroid of that polygon on a dataset that has attributes, we see that R package **sf** issues a warning:

```
# Warning: st_centroid assumes attributes are constant over
# geometries
```

The reason for this is that the dataset contains variables with values that are associated with entire polygons – in this case population counts – meaning they are not associated with a `POINT` geometry replacing the polygon.

In Section 1.6 we already described that for non-point geometries (lines, polygons), feature attribute values either have *point support*, meaning that the value applies to *every point*, or they have *block support*, meaning that the value *summarises all points* in the geometry. (Other options, non-point support smaller than the geometry, or support larger than the associated geometry, may also occur.) This chapter will describe different ways in which an attribute may relate to the geometry, its consequences on analysing such data, and ways to derive attribute data for different geometries (up- and downscaling).

5.1   Attribute-geometry relationships and support

Changing the feature geometry without changing the feature attributes does change the *feature*, since the feature is characterised by the combination of geometry and attributes. Can we, ahead of time, predict whether the resulting feature will still meaningfully relate to the attribute value when we replace all geometries for instance with their convex hull or centroid? It depends.

Take the example of a road, represented by a `LINESTRING`, which has an attribute property *road width* equal to 10 m. What can we say about the road width of an arbitrary sub-section of this road? That depends on whether the attribute road length describes, for instance the road width *everywhere*, meaning that road width is constant along the road, or whether it describes an aggregate property, such as minimum or average road width. In case of the minimum, for an arbitrary subsection of the road one could still argue that the minimum road width must be at least as large as the minimum road width for the whole segment, but it may no longer be *the minimum* for that subsection. This gives us two "types" for the attribute-geometry relationship (**AGR**):

- **constant** the attribute value is valid everywhere in or over the geometry; we can think of the feature as consisting of an infinite number of points that all have this attribute value; in geostatistical terminology this is known as a variable with *point support*
- **aggregate** the attribute is an aggregate, a summary value over the geometry; we can think of the feature as a *single* observation with a value that is associated with the *entire* geometry; this is also known as a variable having *block support*

For polygon data, typical examples of **constant** AGR (point support) variables are:

- land use for a land use polygon
- rock units or geologic strata in a geological map
- soil type in a soil map
- elevation class in an elevation map that shows elevation as classes
- climate zone in a climate zone map

A typical property of such variables is that they have geometries that are not man-made and also not associated with a sensor device (such as remote sensing image pixel boundaries). Instead, the geometry follows from mapping the variable observed.

Examples for the **aggregate** AGR (block support) variables are:

- population, either as number of persons or as population density
- other socio-economic variables, summarised by area
- average reflectance over a remote sensing pixel
- total emission of pollutants by region
- block mean NO_2 concentrations, such as obtained by block kriging over square blocks (Section 12.5) or by a dispersion model that predicts areal means

A typical property of such variables is that associated geometries come for instance from legislation, observation devices or analysis choices, but not intrinsically from the observed variable.

A third type of AGR arises when an attribute *identifies* a feature geometry; we call an attribute an **identity** variable when the associated geometry uniquely identifies the variable's value (there are no other geometries with the same value). An example is county name: the name identifies the county, and is still the county for any sub-area (point support), but for arbitrary sub-areas, the attributes loses the **identity** property to become a **constant** attribute. An example is:

- an arbitrary point (or region) inside a county is still part of the county and must have the same value for county name, but it no longer identifies the (entire) geometry corresponding to that county

The challenge here is that spatial information (ignoring time for simplicity) belongs to different phenomena types (Scheider et al. 2016), including:

- **fields**: where over *continuous* space, every location corresponds to a single value, examples including elevation, air quality, or land use
- **objects**: found at a *discrete* set of locations, such as houses, trees, or persons
- **aggregates**: values arising as spatial sums, totals, or averages of fields, counts or densities of objects, associated with lines or regions

but that different spatial geometry types (points, lines, polygons, raster cells) have no simple mapping to these phenomena types:

- points may refer to sample locations of observations on fields (air quality) or to locations of objects
- lines may be used for objects (roads, rivers), contours of a field, or administrative borders
- raster pixels and polygons may reflect fields of a categorical variable such as land use (*coverage*), but also aggregates such as population density
- raster or other mesh triangulations may have different variables associated with nodes (points), edges (lines), or faces (areas, cells), for instance when partial differential equations are approximated using *staggered grids* (Haltiner and Williams 1980; Collins et al. 2013)

Properly specifying attribute-geometry relationships, and warning against their absence or cases when change in geometry (change of support) implies a change of information can help to avoid a large class of common spatial data analysis mistakes (Stasch et al. 2014) associated with the *support* of spatial data.

5.2   Aggregating and summarising

Aggregating records in a table (or `data.frame`) involves two steps:

- grouping records based on a grouping predicate, and
- applying an aggregation function to the attribute values of a group to summarise them into a single number.

In SQL, this looks for instance like

`SELECT GroupID, SUM(population) FROM table GROUP BY GroupID;`

indicating the aggregation *function* (`SUM`) and the *grouping predicate* (`GroupID`).

R package **dplyr** for instance uses two steps to accomplish this: function `group_by` specifies the group membership of records, `summarise` computes data summaries (such as `sum` or `mean`) for each of the groups. The (base) R method `aggregate` combines both in a single function call that takes the table, the grouping predicate(s), and the aggregation function(s) as arguments.

An example for the North Carolina counties is shown in Figure 5.1. Here, we grouped counties by their position (according to the quadrant in which the county centroid is with respect to ellipsoidal coordinate `POINT(-79, 35.5)`) and summed the number of disease cases per group. The result shows that the geometries of the resulting groups have been unioned (Section 3.2.6): this is necessary because the `MULTIPOLYGON` formed by just putting all the county geometries together would have many duplicate boundaries, and hence would not be *valid* (Section 3.1.2).

Plotting collated county polygons is technically not a problem, but for this case would raise the wrong suggestion that the group sums relate to individual counties, rather than to the grouped counties.

One particular property of aggregation in this way is that each record is assigned to a single group; this has the advantage that the sum of the group-wise sums equals the sum of the un-grouped data: for variables that reflect *amount*, nothing gets lost and nothing is added. The newly formed geometry is the result of unioning the geometries of the contributing records.

When we need an aggregate for a new area that is *not* a union of the geometries for a group of records, and we use a spatial predicate, then single records may be matched to multiple groups. When taking the rectangles of Figure 5.2 as the target areas, and summing for each rectangle the disease cases of the counties that *intersect* with the rectangles of Figure 5.2, the sum of these will be much larger:

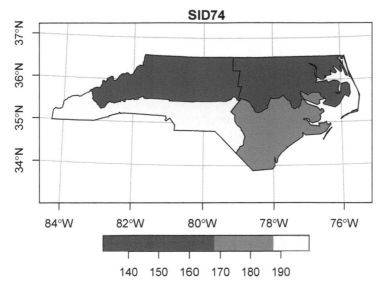

Figure 5.1: SID74 total incidences aggregated to four areas

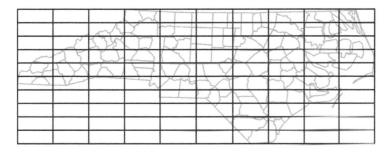

Figure 5.2: Example target blocks plotted over North Carolina counties

```
#    sid74_sum_counties sid74_sum_rectangles
#                667                 2621
```

Choosing another predicate, such as *contains* or *covers*, would on the contrary result in much smaller values, because many counties are not contained by *any* of the target geometries. However, there are a few cases where this approach might be good or satisfactory:

- when we want to aggregate POINT geometries by a set of polygons, and all points are contained by a single polygon. If points fall on a shared boundary than they are assigned to both polygons (this is the case for DE-9IM-based GEOS library; the s2geometry library has the option to define polygons as "semi-open", which implies that points are assigned to maximally one polygons when polygons do not overlap)

- when aggregating many very small polygons or raster pixels over larger areas, for instance *averaging* altitude from a 30 m resolution raster over North Carolina counties, the error made by multiple matches may be insignificant
- when the many-to-many match is reduced to the single largest area match, as shown in Figure 7.4

A more comprehensive approach to aggregating spatial data associated with areas to larger, arbitrary shaped areas is by using area-weighted interpolation.

5.3   Area-weighted interpolation

When we want to combine geometries and attributes of two datasets such that we get attribute values of a source dataset summarised for the geometries of a target, where source and target geometries are unrelated, area-weighted interpolation may be a simple approach. In effect, it considers the area of overlap of the source and target geometries, and uses that to weigh the source attribute values into the target value (Goodchild and Lam 1980; Do, Thomas-Agnan, and Vanhems 2015a, 2015b; Do, Laurent, and Vanhems 2021). This methods is also known as conservative region aggregation or regridding (Jones 1999). Here, we follow the notation of Do, Thomas-Agnan, and Vanhems (2015b).

Area-weighted interpolation computes, for each of q spatial target areas T_j, a weighted average from the values Y_i corresponding to the p spatial source areas S_i,

$$\hat{Y}_j(T_j) = \sum_{i=1}^{p} w_{ij} Y_i(S_i) \tag{5.1}$$

where the w_{ij} depend on the amount of overlap of T_j and S_i, and the amount of overlap is $A_{ij} = T_j \cap S_i$. How w_{ij} depends on A_{ij} is discussed below.

Different options exist for choosing weights, including methods using external variables (including dasymetric mapping, Mennis 2003). Two simple approaches for computing weights that do not use external variables arise, depending on whether the variable Y is *intensive* or *extensive*.

5.3.1   Spatially extensive and intensive variables

Extensive variables correspond to amounts, associated with a physical size (length, area, volume, counts of items). An example of a extensive variable is *population count*. It is associated with an area, and if that area is cut into smaller areas, the population count needs to be split too. Because population is rarely uniform over space, this does not need to be done proportionally to the smaller areas but the sum of the population count for the smaller areas needs to equal that of the total. Intensive variables are variables that do not have values proportional to support: if the area is split, values may vary but *on average* remain the same. An example of a related variable that is *intensive* is population density. If an area is split into smaller areas, population density is not split similarly: the *sum* of population densities for the smaller areas is a meaningless measure, as opposed to the *average* of the population densities which will be similar to the density of the total area.

When we assume that the **extensive** variable Y is uniformly distributed over space, the value Y_{ij}, derived from Y_i for a sub-area of S_i, $A_{ij} = T_j \cap S_i$ of S_i is

$$\hat{Y}_{ij}(A_{ij}) = \frac{|A_{ij}|}{|S_i|} Y_i(S_i)$$

where $|\cdot|$ denotes the spatial area. For estimating $Y_j(T_j)$ we sum all the elements over area T_j:

$$\hat{Y}_j(T_j) = \sum_{i=1}^{p} \frac{|A_{ij}|}{|S_i|} Y_i(S_i) \tag{5.2}$$

For an **intensive** variable, under the assumption that the variable has a constant value over each area S_i, the estimate for a sub-area equals that of the total area,

$$\hat{Y}_{ij} = Y_i(S_i)$$

and we can estimate the value of Y for a new spatial unit T_j by an area-weighted average of the source values:

$$\hat{Y}_j(T_j) = \sum_{i=1}^{p} \frac{|A_{ij}|}{|T_j|} Y_i(S_i) \tag{5.3}$$

5.3.2 Dasymetric mapping

Dasymetric mapping distributes variables, such as population, known at a coarse spatial aggregation level over finer spatial units by using other variables that are associated with population distribution, such as land use, building density, or road density. The simplest approach to dasymetric mapping is obtained for extensive variables, where the ratio $|A_{ij}|/|S_i|$ in Equation 5.2 is replaced by the ratio of another extensive variable $X_{ij}(S_{ij})/X_i(S_i)$, which has to be known for both the intersecting regions S_{ij} and the source regions S_i. Do, Thomas-Agnan, and Vanhems (2015b) discuss several alternatives for intensive Y and/or X, and cases where X is known for other areas.

5.3.3 Support in file formats

GDAL's vector API supports reading and writing so-called field domains, which can have a "split policy" and a "merge policy" indicating what should be done with attribute variables when geometries are split or merged. The values of these can be "duplicate" for split and "geometry weighted" for merge, in case of spatially intensive variables, or they can be "geometry ratio" for split and "sum" for merge, in case of spatially extensive variables. At the time of writing this, the file formats supporting this are GeoPackage and FileGDB.

5.4   Up- and Downscaling

Up- and downscaling refers in general to obtaining high-resolution information from low-resolution data (downscaling) or obtaining low-resolution information from high-resolution data (upscaling). Both activities involve the attributes' relation to geometries and both change support. They are synonymous with aggregation (upscaling) and disaggregation (downscaling). The simplest form of downscaling is sampling (or extracting) polygon, line or grid cell values at point locations. This works well for variables with point-support ("constant" AGR), but is at best approximate when the values are aggregates. Challenging applications for downscaling include high-resolution prediction of variables obtained by low-resolution weather prediction models or climate change models, and the high-resolution prediction of satellite image derived variables based on the fusion of sensors with different spatial and temporal resolutions.

The application of areal interpolation using (Equation 5.1) with its realisations for extensive (Equation 5.2) and intensive (Equation 5.3) variables allows moving information from any source area S_i to any target area T_j as long as the two areas have some overlap. This means that one can go arbitrarily to much larger units (aggregation) or to much smaller units (disaggregation). Of course this makes only sense to the extent that the assumptions hold: over the source regions extensive variables need to be uniformly distributed and intensive variables need to have a constant value.

The ultimate disaggregation involves retrieving (extracting) point values from line or area data. For this, we cannot work with equations -Equation 5.2 or -Equation 5.3 because $|A_{ij}| = 0$ for points, but under the assumption of having a constant value over the geometry, for intensive variables the value $Y_i(S_i)$ can be assigned to points as long as all points can be uniquely assigned to a single source area S_i. For polygon data, this implies that Y needs to be a coverage variable (Section 3.4). For extensive variables, extracting a value at a point is rather meaningless, as it should always give a zero value.

In cases where values associated with areas are **aggregate** values over the area, the assumptions made by area-weighted interpolation or dasymetric mapping – uniformity or constant values over the source areas – are highly unrealistic. In such cases, these simple approaches could still be reasonable approximations, for instance when:

- the source and target area are nearly identical
- the variability inside source units is very small, and the variable is nearly uniform or constant

In other cases, results obtained using these methods are merely consequences of unjustified assumptions. Statistical aggregation methods that can estimate quantities for larger regions from points or smaller regions include:

- design-based methods, which require that a probability sample is available from the target region, with known inclusion probabilities (Brus 2021a, Section 10.4), and

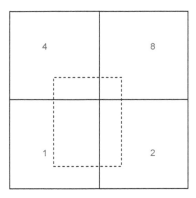

Figure 5.3: Example data for area-weighted interpolation

- model-based methods, which assume a random field model with spatially correlated values (block kriging, Section 12.5)

Alternative disaggregation methods include:

- deterministic, smoothing-based approaches such as kernel- or spline-based smoothing methods (Tobler 1979; Martin 1989)
- statistical, model-based approaches such as area-to-area and area-to-point kriging (Kyriakidis 2004; Raim et al. 2021).

5.5 Exercises

Where relevant, try to make the following exercises with R.

1. When we add a variable to the nc dataset by nc$State = "North Carolina" (i.e., all counties get assigned the same state name). Which value would you attach to this variable for the attribute-geometry relationship (agr)?
2. Create a new sf object from the geometry obtained by st_union(nc), and assign "North Carolina" to the variable State. Which agr can you now assign to this attribute variable?
3. Use st_area to add a variable with name area to nc. Compare the area and AREA variables in the nc dataset. What are the units of AREA? Are the two linearly related? If there are discrepancies, what could be the cause?
4. Is the area variable intensive or extensive? Is its agr equal to constant, identity or aggregate?
5. Consider Figure 5.3: using the equations in Section 5.3.1, compute the area-weighted interpolations for (a) the dashed cell and (b) for the square enclosing all four solid cells, first for the case where the four cells represent (i) an extensive variable, and (ii) an intensive variable. The red numbers are the data values of the source areas.

6

Data Cubes

Data cubes arise naturally when we observe properties of a set of geometries repeatedly over time. Time information may sometimes be considered as an attribute of a feature, such as when we register the year of construction of a building or the date of birth of a person (Chapter 5). In other cases it may refer to the time of observing an attribute, or the time for which a prediction of an attribute has been made. In these cases, time is on equal footing with space, and time and space together describe the physical dimensions over which we observe, model, predict, or make forecasts for.

One way of considering our world is that of a four-dimensional space, with three space dimensions and one time dimension. In that view, events become "things" or "objects" that have as duration their size on the time dimension (Galton 2004). Although such a view does not align well with how we experience and describe the world, from a data analytical perspective, four numbers, along with their reference systems, suffice to describe space and time coordinates of an observation associated with a point location and time instance.

We define data cubes as array data with one or more array dimensions associated with space and/or time (Lu, Appel, and Pebesma 2018). This implies that raster data, features with attributes, and time series data are all special cases of data cubes. Since we do not restrict to three-dimensional structures, we actually mean *hypercubes* rather than cubes, and as the cube extent of the different dimensions does not have to be identical, or have comparable units, the better term would be *hyper-rectangle*. For simplicity, we talk about data cubes instead.

A canonical form of a data cube is shown in Figure 6.1: it shows in a perspective plot a set of raster layers for the same region that were collected (observed, or modelled) at different time steps. The three cube dimensions longitude, latitude, and time, are thought of as being orthogonal. Arbitrary two-dimensional cube slices are obtained by fixing one of the dimensions at a particular value, one-dimensional slices are obtained by fixing two of the dimensions at a particular value, and a scalar is obtained by fixing three dimensions at a particular value.

6.1 A four-dimensional data cube

Figure 6.2 depicts a four-dimensional raster data cube (Appel and Pebesma 2019), where three-dimensional raster data cubes with a spectral dimension ("bands") are organised along a fourth dimension, a time axis. Colour image data always has three

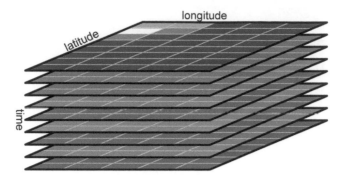

Figure 6.1: Raster data cube with dimensions latitude, longitude, and time

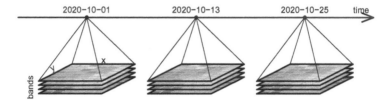

Figure 6.2: Four-dimensional raster data cube with dimensions x, y, bands, and time

bands (blue, green, red), and this example has a fourth band (near infrared, NIR), which is commonly found in spectral remote sensing data.

Figure 6.3 shows exactly the same data, but layed out flat as a facet plot (or scatterplot matrix), where two dimensions (x and y) are aligned with (or nested within) the dimensions *bands* and *time*, respectively.

6.2   Dimensions, attributes, and support

Phenomena in space and time can be thought of as functions with *domain* space and time, and with *range* one or more observed attributes. For clearly identifiable discrete events or objects, the range is typically discrete, and precise delineation involves describing the precise coordinates where a thing starts or stops, which is best suited by vector geometries. For continuous phenomena, variables that take on a value everywhere such as air temperature or land use type, there are infinitely many values to represent and a common approach is to discretise space and time *regularly* over the spatiotemporal domain (extent) of interest. This leads to a number of familiar data structures:

- time series, depicted as time lines for functions of time
- image or raster data for two-dimensional spatial data

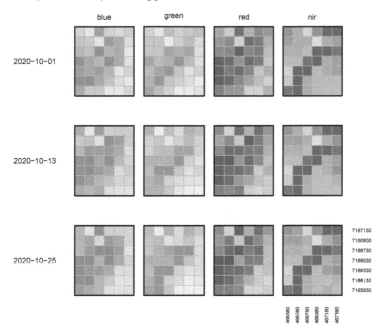

Figure 6.3: Four-dimensional raster data cube layed out flat over two dimensions

- time sequences of images for dynamic spatial data

The third form of this, where a variable Z depends on x, y and t, as in

$$Z = f(x, y, t)$$

is the archetype of a spatiotemporal array or *data cube*: the shape of the volume where points regularly discretising the domain forms a cube. We call the variables that form the range (here: x, y, t) the cube *dimensions*. Data cubes may have multiple attributes, as in

$$\{Z_1, Z_2, ..., Z_p\} = f(x, y, t)$$

and if Z is functional, for instance reflectance values measured over the electromagnetic spectrum, the spectral wavelengths λ may form an additional dimension, as in $Z = f(x, y, t, \lambda)$. Section 6.5 discusses the alternative of representing colour bands as attributes.

Multiple time dimensions arise for instance when making forecasts for different times in the future t' at different times t, or when time is split over multiple dimensions (as in year, day-of-year, and hour-of-day). The most general definition of a data cube is a functional mapping from n dimensions to p attributes:

$$\{Z_1, Z_2, ..., Z_p\} = f(D_1, D_2, ..., D_n)$$

Here, we will consider any dataset with one or more space dimensions and zero or more time dimensions as data cubes. That includes:

- simple features (Section 3.1)
- time series for sets of features
- raster data
- multi-spectral raster data (images)
- time series of multi-spectral raster data (video)

6.2.1   Regular dimensions, GDAL's geotransform

Data cubes are usually stored in multi-dimensional arrays, and the usual relationship between 1-based array index i and an associated regularly discretised dimension variable x is

$$x = o_x + (i - 1)d_x$$

with o_x the origin, and d_x the grid spacing for this dimension.

For more general cases like those in Figures -Figure 1.6 b-c, the relation between x and y and array indexes i and j is

$$x = o_x + (i - 1)d_x + (j - 1)a_1$$

$$y = o_y + (i - 1)a_2 + (j - 1)d_y$$

With two affine parameters a_1 and a_2; this is the so-called *geotransform* as used in GDAL. When $a_1 = a_2 = 0$, this reduces to the regular raster of Figure 1.6 a with square cells if $d_x = d_y$. For integer indexes, the coordinates are that of the starting *edge* of a grid cell, and the cell area (pixel) spans a range corresponding to index values ranging from i (inclusive) to $i + 1$ (exclusive). For most common imagery formats, d_y is negative, indicating that image row index increases with decreasing y values (southward). To get the x- and y-coordinate of the grid cell *centre* of the top left grid cell (in case of a negative d_y), we use $i = 1.5$ and $j = 1.5$.

For rectilinear rasters, a table that maps array index to dimension values is needed. NetCDF files for instance always store all values of spatial dimension (coordinate) variables that may correspond to the centre or offset of spatial grid cells, and in addition may store grid cell boundaries (to define rectilinear dimensions or to disambiguate the relation of the coordinate variable values to the cell boundaries).

For curvilinear rasters, an array that maps every combination of i, j into x, y pairs is needed, or a parametric function that does this (a projection or its inverse). NetCDF files often provide both, when available. GDAL calls such arrays *geolocation arrays*, and has extensive support for transformations involving them.

6.2.2   Support along cube dimensions

Section 5.1 defined *spatial* support of an attribute variable as the size (length, area, volume) of a geometry associated with a particular observation or prediction. The same notion applies to temporal support. Although time is rarely reported by explicit time periods having a start- and end-time, in many cases either the time stamp implies a period (ISO-8601 uses "2021" for the full year, "2021-01" for the full month) or the time period is taken as the period from the time stamp of the current record up to but not including the time stamp of the next record.

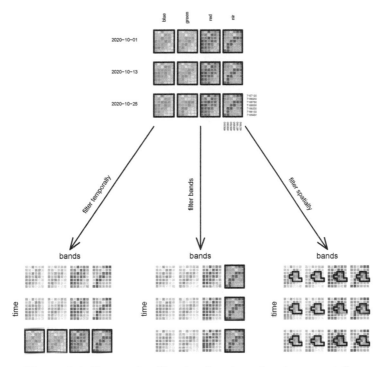

Figure 6.4: Data cube filtering by time, band or spatially

An example is MODIS satellite imagery, where vegetation indexes (NDVI and EVI) are available as 16-day composites, meaning that over 16-day periods all available imagery is aggregated into a single image; such composites have temporal "block support". Sentinel-2 or Landsat-8 data on the other hand are "snapshot" images and have temporal "point support". When temporally aggregating data with temporal point support, for instance to monthly values, all images falling in the target time interval are selected. When aggregating temporal block support imagery such as the MODIS 16-day composite, one might weigh images according to the amount of overlap of the 16-day composite period and the target period, similar to area-weighted interpolation Section 5.3 but over the time dimension.

6.3 Operations on data cubes

6.3.1 Slicing a cube: filter

Data cubes can be sliced into sub-cubes by fixing a dimension at a particular value. Figure 6.4 shows the sub-cubes obtained by doing this with the temporal, spectral, or spatial dimension. In this figure, the spatial filtering does not happen by fixing a single spatial dimension at a particular value, but by selecting a particular sub-region,

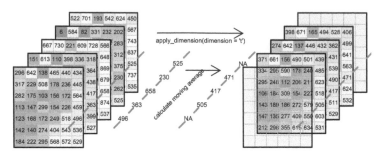

Figure 6.5: Low-pass filtering of time series

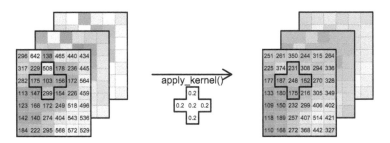

Figure 6.6: Low pass filtering of spatial slices

which is a more common operation. Fixing x or y would give a sub-cube along a transect of constant x or y, which can be used to show a Hovmöller diagram, where an attribute is plotted (coloured) in the space of one space and one time dimension.

6.3.2   Applying functions to dimensions

A common analysis involves applying a function over one or more cube dimensions. Simple cases arise where a function such as **abs**, **sin**, or **sqrt** is applied to all values in the cube, or when a function takes all values in the cube and returns a single scalar, such as when computing the mean or maximum value over the entire cube. Other options include applying the function to selected dimensions, such as applying a *temporal* low-pass filter to every individual (pixel/band) time series as shown in Figure 6.5, or applying a *spatial* low-pass filter to every spatial slice for every band/time combination, shown in Figure 6.6.

6.3.3   Reducing dimensions

When applying function **mean** to an entire data cube, all dimensions vanish: the resulting "data cube" has dimensionality zero. We can also apply functions to a limited set of dimensions such that selected dimensions vanish, or are *reduced*. We already saw that filtering is a special case of this, but more in general we could for instance compute the maximum of every time series, the mean over every spatial slice, or a band index such as NDVI that summarises different spectral values into a single new "band" with the index value. Figure 6.7 illustrates these options.

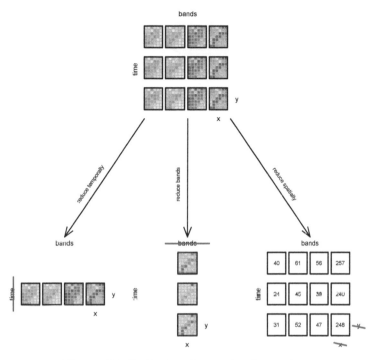

Figure 6.7: Reducing data cube dimensions

6.4 Aggregating raster to vector cubes

Figure 6.8 illustrates how a four-dimensional raster data cube can be aggregated to a three-dimensional *vector data cube*. Pixels in the raster are grouped by spatial intersection with a set of vector geometries, and each group is then reduced to a single value by an aggregation function such as `mean` or `max`. In the example, the *two* spatial dimensions x and y reduce to a single dimension, the one-dimensional sequence of feature geometries, with geometries that are defined in the space of x and y. Grouping geometries can also be `POINT` geometries, in which case the aggregation function is obsolete as single values at the `POINT` locations are *extracted* by querying a pixel value or by interpolating from the nearest pixels.

Further examples of vector data cubes include air quality data, where we could have PM_{10} measurements over two dimensions, as a sequence of

- monitoring stations, and
- time intervals

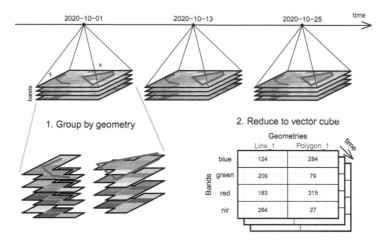

Figure 6.8: Aggregating a raster data cube to a vector data cube

or where we consider time series of demographic or epidemiological data, consisting of (population, disease) counts, with number of persons by

- region, for a sequence of n regions
- age class, for m age classes, and
- year, for p years

which forms an array with nmp elements.

For spatial data science, handling vector and raster data cubes is extremely useful, because many variables are both spatially and temporaly varying, and because we often want to either change dimensions or aggregate them out, but in a fully flexible manner and order. Examples of changing dimensions are:

- interpolating air quality measurements to values on a regular grid (raster; Chapter 12)
- estimating density maps from points or lines, for instance estimating the number of flights passing by per week within a range of 1 km (Chapter 11)
- aggregating climate model predictions to summary indicators for administrative regions
- combining Earth observation data from different sensors, such as MODIS (250~m pixels, every 16 days) with Sentinel-2 (10~m pixels, every 5 days)

Examples of aggregating one or more full dimensions are assessments of:

- which air quality monitoring stations indicate unhealthy conditions (time)
- which region has the highest increase in disease incidence (space, time)
- global warming (global change in degrees Celcius per decade)

6.5 Switching dimension with attributes

When we accept that a dimension can also reflect an unordered, categorical variable, then one can easily swap a set of attributes for a single dimension, by replacing

$$\{Z_1, Z_2, ..., Z_p\} = f(D_1, D_2, ..., D_n)$$

with

$$Z = f(D_1, D_2, ..., D_n, D_{n+1})$$

where D_{n+1} has cardinality p and has as labels (the names of) $Z_1, Z_2, ..., Z_p$. Figure 6.9 shows a vector data cube for air quality stations where one cube dimension reflects air quality parameters. When the Z_i have incompatible measurement units, as in Figure 6.9, one would have to take care when reducing the "parameter" dimension D_{n+1}: numeric functions like `mean` or `max` would be meaningless. Counting the number of variables that exceed their respective threshold values may however be meaningful.

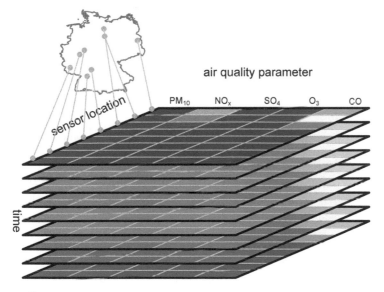

Figure 6.9: Vector data cube with air quality time series

Being able to swap dimensions to attributes flexibly and vice versa leads to highly flexible analysis possibilities (P. G. Brown 2010).

6.6 Other dynamic spatial data

We have seen several dynamic raster and vector data examples that match the data cube structure well. Other data examples do less so: in particular spatiotemporal point patterns (Chapter 11) and trajectories [movement data; for a recent review, see Joo et al. (2020)] are often more straightforward to not handle as a data cube. Spatiotemporal point patterns are the sets of spatiotemporal coordinates of events or objects: accidents, disease cases, traffic jams, lightning strikes, and so on. Trajectory data are time sequences of spatial locations of moving objects (persons, cars, satellites, animals). For such data, the primary information is in the coordinates, and shifting these to a limited set of regularly discretised grid cells covering the space may help some analysis, for instance to quickly explore patterns in areas of higher densities, but the loss of the exact coordinates also hinders a number of analysis approaches involving distance, direction, or speed calculations. Nevertheless, for such data often the first computational steps involves generation of data cube representations by aggregating to a time-fixed spatial and/or space-fixed temporal discretisation.

Using sparse array representations of data cubes to represent point pattern or trajectory data, which is possible for instance with SciDB (P. G. Brown 2010) or TileDB (Papadopoulos et al. 2016), may strongly limit the loss of coordinate accuracy by choosing dimensions that represent an extremely dense grid, and storing only those grid cells that contain data points. For trajectory data, such representations would need to add a grouping dimension to identify individuals, or individual sequences of consecutive movement observations.

6.7   Exercises

Use words to solve the following exercises. If needed or relevant use R code to illustrate the argument(s).

1. Why is it difficult to represent trajectories, sequences of (x, y, t) obtained by tracking moving objects, by data cubes as described in this chapter?
2. In a socio-economic vector data cube with variables population, life expectancy, and gross domestic product ordered by dimensions country and year, which variables have block support for the spatial dimension, and which have block support for the temporal dimension?
3. The Sentinel-2 satellites collect images in 12 spectral bands; list advantages and disadvantages to represent them as (i) different data cubes, (ii) a data cube with 12 attributes, one for each band, and (iii) a single attribute data cube with a spectral dimension.
4. Explain why a curvilinear raster as shown in Figure 1.6 can be considered a special case of a data cube.
5. Explain how the following problems can be solved with data cube operations **filter**, **apply**, **reduce** and/or **aggregate**, and in which order. Also mention for each which function is applied, and what the dimensionality of the resulting data cube is (if any):
    * from hourly PM_{10} measurements for a set of air quality monitoring stations, compute per station the amount of days per year that the average daily PM_{10} value exceeds 50 $\mu g/m^3$

- for a sequence of aerial images of an oil spill, find the time at which the oil spill had its largest extent, and the corresponding extent
- from a 10-year period with global daily sea surface temperature (SST) raster maps, find the area with the 10% largest and 10% smallest temporal trends in SST values.

Part II

R for Spatial Data Science

The second part of this book explains how the concepts introduced in the first part are dealt with using R. Chapter 7 deals with basic handling of spatial data: reading, writing, subsetting, selecting by spatial predicates, geometry transformers like buffers or intersections, raster-vector and vector-raster conversion, handling of data cubes, spherical geometry, coordinate transformations and conversions. This is followed by Chapter 8 which is dedicated to plotting of spatial and spatiotemporal data with base plot, and packages **ggplot2**, **tmap** and **mapview**. The chapter deals with projection, colours, colour breaks, graticules, graphic elements on maps like legends, and interactive maps. Chapter 9 discusses approaches to handle large vector or raster datasets or data cubes, where "large" either means too large to fit in memory or too large to download.

The material covered in this part is not meant as a complete tutorial nor a manual of the packages covered, but rather as an explanation and illustration of a number of common workflows. More complete and detailed information is found in the package documentation, in particular in the package vignettes for packages **sf** and **stars**. Links to them are found on the CRAN landing pages of the packages.

7

Introduction to sf and stars

This chapter introduces R packages **sf** and **stars**. **sf** provides a table format for simple features, where feature geometries are stored in a list-column. R package **stars** was written to support raster and vector data cubes (Chapter 6), supporting raster layers, raster stacks and feature time series as special cases. **sf** first appeared on CRAN in 2016, **stars** in 2018. Development of both packages received support from the R Consortium as well as strong community engagement. The packages were designed to work together. Functions or methods operating on **sf** or **stars** objects start with **st_**, making it easy to recognise them or to search for them when using command line completion.

7.1 Package sf

Intended to succeed and replace R packages **sp**, **rgeos** and the vector parts of **rgdal**, R package **sf** (Pebesma 2018) was developed to move spatial data analysis in R closer to standards-based approaches seen in the industry and open source projects, to build upon more modern versions of the open source geospatial software stack (Figure 1.7), and to allow for integration of R spatial software with the tidyverse (Wickham et al. 2019), if desired.

To do so, R package **sf** provides simple features access (Herring et al. 2011), natively, to R. It provides an interface to several **tidyverse** packages, in particular to **ggplot2**, **dplyr**, and **tidyr**. It can read and write data through GDAL, execute geometrical operations using GEOS (for projected coordinates) or s2geometry (for ellipsoidal coordinates), and carry out coordinate transformations or conversions using PROJ. External C++ libraries are interfaced using R package **Rcpp** (Eddelbuettel 2013).

Package **sf** represents sets of simple features in **sf** objects, a sub-class of a `data.frame` or tibble. **sf** objects contain at least one *geometry list-column* of class `sfc`, which for each element contains the geometry as an R object of class `sfg`. A geometry list-column acts as a variable in a `data.frame` or tibble, but has a more complex structure than basic vectors such as numeric or character variables Section B.3.

An **sf** object has the following metadata:

- the name of the (active) geometry column, held in attribute `sf_column`
- for each non-geometry variable, the attribute-geometry relationship (Section 5.1), held in attribute `agr`

75

An `sf` geometry list-column is extracted from an `sf` object with `st_geometry` and has the following metadata:

- coordinate reference system held in attribute `crs`
- bounding box held in attribute `bbox`
- precision held in attribute `precision`
- number of empty geometries held in attribute `n_empty`

These attributes may best be accessed or set by using functions like `st_bbox`, `st_crs`, `st_set_crs`, `st_agr`, `st_set_agr`, `st_precision`, and `st_set_precision`.

Geometry columns in `sf` objects can be set or replaced using `st_geometry<-` or `st_set_geometry`.

7.1.1   Creation

An `sf` object can be created from scratch by

```
library(sf)
# Linking to GEOS 3.11.1, GDAL 3.6.2, PROJ 9.1.1; sf_use_s2()
# is TRUE
p1 <- st_point(c(7.35, 52.42))
p2 <- st_point(c(7.22, 52.18))
p3 <- st_point(c(7.44, 52.19))
sfc <- st_sfc(list(p1, p2, p3), crs = 'OGC:CRS84')
st_sf(elev = c(33.2, 52.1, 81.2),
      marker = c("Id01", "Id02", "Id03"), geom = sfc)
# Simple feature collection with 3 features and 2 fields
# Geometry type: POINT
# Dimension:     XY
# Bounding box:  xmin: 7.22 ymin: 52.2 xmax: 7.44 ymax: 52.4
# Geodetic CRS:  WGS 84
#   elev marker                 geom
# 1 33.2   Id01 POINT (7.35 52.4)
# 2 52.1   Id02 POINT (7.22 52.2)
# 3 81.2   Id03 POINT (7.44 52.2)
```

Figure 7.1 gives an explanation of the components printed. Rather than creating objects from scratch, spatial data in R are typically read from an external source, which can be:

- external file
- table (or set of tables) in a database
- request to a web service
- dataset held in some form in another R package

The next section introduces reading from files. Section 9.1 discusses handling of datasets too large to fit into working memory.

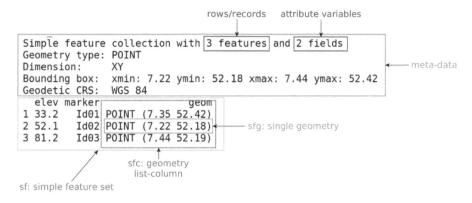

Figure 7.1: components of an **sf** object

7.1.2 Reading and writing

Reading datasets from an external "data source" (file, web service, or even string) is done using **st_read**:

```
library(sf)
(file <- system.file("gpkg/nc.gpkg", package = "sf"))
# [1] "/home/edzer/R/x86_64-pc-linux-gnu-library/4.0/sf/gpkg/nc.gpkg"
nc <- st_read(file)
# Reading layer `nc.gpkg' from data source
#   `/home/edzer/R/x86_64-pc-linux-gnu-library/4.0/sf/gpkg/nc.gpkg'
#   using driver `GPKG'
# Simple feature collection with 100 features and 14 fields
# Geometry type: MULTIPOLYGON
# Dimension:     XY
# Bounding box:  xmin: -84.3 ymin: 33.9 xmax: -75.5 ymax: 36.6
# Geodetic CRS:  NAD27
```

Here, the file name and path **file** is read from the **sf** package, which has a different path on every machine, and hence is guaranteed to be present on every **sf** installation.

Command **st_read** has two arguments: the *data source name* (**dsn**) and the *layer*. In the example above, the *geopackage* (GPKG) file contains only a single layer that is being read. If it had contained multiple layers, then the first layer would have been read and a warning would have been emitted. The available layers of a dataset can be queried by

```
st_layers(file)
# Driver: GPKG
# Available layers:
```

```
#   layer_name geometry_type features fields crs_name
# 1    nc.gpkg Multi Polygon      100     14    NAD27
```

Simple feature objects can be written with **st_write**, as in

```
(file = tempfile(fileext = ".gpkg"))
# [1] "/tmp/Rtmp4s4ZLr/file2857502554643.gpkg"
st_write(nc, file, layer = "layer_nc")
# Writing layer `layer_nc' to data source
#   `/tmp/Rtmp4s4ZLr/file2857502554643.gpkg' using driver `GPKG'
# Writing 100 features with 14 fields and geometry type Multi Polygon.
```

where the file format (GPKG) is derived from the file name extension. Using argument
append, **st_write** can either append records to an existing layer or replace it; if unset,
it will err if a layer already exists. The tidyverse-style **write_sf** will replace silently
if **append** has not been set. Layers can also be deleted using **st_delete**, which is
convenient in particular when they are associated with tables in a database.

For file formats supporting a WKT2 coordinate reference system, **sf_read** and
sf_write will read and write it. For simple formats such as **csv** this will not work.
The shapefile format supports only a very limited encoding of the CRS.

7.1.3   Subsetting

A very common operation is to subset objects; base R can use [for this. The rules
that apply to **data.frame** objects also apply to **sf** objects: records 2-5 and columns
3-7 are selected by

```
nc[2:5, 3:7]
```

but with a few additional features, in particular:

- the **drop** argument is by default **FALSE** meaning that the geometry column is *always*
  selected, and an **sf** object is returned; when it is set to **TRUE** and the geometry
  column is *not* selected, it is dropped and a **data.frame** is returned
- selection with a spatial (**sf**, **sfc** or **sfg**) object as first argument leads to selection
  of the features that spatially *intersect* with that object (see next section); other
  predicates than *intersects* can be chosen by setting parameter **op** to a function such
  as **st_covers** or any other binary predicate function listed in Section 3.2.2.

7.1.4   Binary predicates

Binary predicates like **st_intersects**, **st_covers** and such (Section 3.2.2) take two
sets of features or feature geometries and return for all pairs whether the predicate is
TRUE or **FALSE**. For large sets this would potentially result in a huge matrix, typically
filled mostly with **FALSE** values and for that reason a sparse representation is returned
by default:

```
nc5 <- nc[1:5, ]
nc7 <- nc[1:7, ]
(i <- st_intersects(nc5, nc7))
# Sparse geometry binary predicate list of length 5,
# where the predicate was `intersects'
#  1: 1, 2
#  2: 1, 2, 3
#  3: 2, 3
#  4: 4, 7
#  5: 5, 6
```

Figure 7.2: First seven North Carolina counties

Figure 7.2 shows how the intersections of the first five with the first seven counties can be understood. We can transform the sparse logical matrix into a dense matrix by

```
as.matrix(i)
#         [,1]   [,2]  [,3]  [,4]  [,5]  [,6]  [,7]
# [1,]   TRUE   TRUE FALSE FALSE FALSE FALSE FALSE
# [2,]   TRUE   TRUE  TRUE FALSE FALSE FALSE FALSE
# [3,] FALSE   TRUE  TRUE FALSE FALSE FALSE FALSE
# [4,] FALSE FALSE FALSE  TRUE FALSE FALSE  TRUE
# [5,] FALSE FALSE FALSE FALSE  TRUE  TRUE FALSE
```

The number of counties that each of **nc5** intersects with is

```
lengths(i)
# [1] 2 3 2 2 2
```

and the other way around, the number of counties in **nc5** that intersect with each of the counties in **nc7** is

```
lengths(t(i))
# [1] 2 3 2 1 1 1 1
```

The object **i** is of class **sgbp** (sparse geometrical binary predicate), and is a list of integer vectors, with each element representing a row in the logical predicate matrix holding the column indices of the **TRUE** values for that row. It further holds some metadata like the predicate used, and the total number of columns. Methods available for **sgbp** objects include

```
methods(class = "sgbp")
# [1] as.data.frame as.matrix    coerce       dim
# [5] initialize    Ops          print        show
# [9] slotsFromS3   t
# see '?methods' for accessing help and source code
```

where the only `Ops` method available is `!`, the negation operation.

7.1.5   tidyverse

The **tidyverse** package loads a collection of data science packages that work well
together (Wickham and Grolemund 2017; Wickham et al. 2019). Package **sf** has
tidyverse-style read and write functions, `read_sf` and `write_sf` that

- return a tibble rather than a `data.frame`,
- do not print any output, and
- overwrite existing data by default.

Further **tidyverse** generics with methods for **sf** objects include `filter`, `select`,
`group_by`, `ungroup`, `mutate`, `transmute`, `rowwise`, `rename`, `slice`, `summarise`,
`distinct`, `gather`, `pivot_longer`, `spread`, `nest`, `unnest`, `unite`, `separate`,
`separate_rows`, `sample_n`, and `sample_frac`. Most of these methods simply manage
the metadata of **sf** objects and make sure the geometry remains present. In case a
user wants the geometry to be removed, one can use `st_drop_geometry` or simply
coerce to a `tibble` or `data.frame` before selecting:

```
library(tidyverse) |> suppressPackageStartupMessages()
nc |> as_tibble() |> select(BIR74) |> head(3)
# # A tibble: 3 x 1
#     BIR74
#     <dbl>
# 1   1091
# 2    487
# 3   3188
```

The **summarise** method for **sf** objects has two special arguments:

- `do_union` (default `TRUE`) determines whether grouped geometries are unioned on
  return, so that they form a valid geometry
- `is_coverage` (default `FALSE`) in case the geometries grouped form a coverage (do
  not have overlaps), setting this to `TRUE` speeds up the unioning

The `distinct` method selects distinct records, where `st_equals` is used to evaluate
distinctness of geometries.

`filter` can be used with the usual predicates; when one wants to use it with a
spatial predicate, for instance to select all counties less than 50 km away from Orange
County, one could use

```
orange <- nc |> dplyr::filter(NAME == "Orange")
wd <- st_is_within_distance(nc, orange,
                            units::set_units(50, km))
o50 <- nc |> dplyr::filter(lengths(wd) > 0)
nrow(o50)
# [1] 17
```

(where we use `dplyr::filter` rather than `filter` to avoid confusion with `filter` from base R.)

Figure 7.3 shows the results of this analysis, and in addition a buffer around the county borders. Note that this buffer serves for illustration: it was *not* used to select the counties.

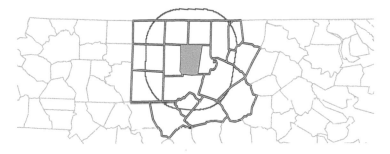

Figure 7.3: Orange County (orange), counties within a 50 km radius (black), a 50~km buffer around Orange County (brown), and remaining counties (grey)

7.2 Spatial joins

In regular (left, right, or inner) joins, *joined* records from a pair of tables are reported when one or more selected attributes match (are identical) in both tables. A spatial join is similar, but the criterion to join records is not equality of attributes but a spatial predicate. This leaves a wide variety of options in order to define *spatially matching records*, using binary predicates listed in Section 3.2.2. The concepts of "left", "right", "inner", or "full" joins remain identical to the non-spatial join as the options for handling records that have no spatial match.

When using spatial joins, each record may have several matched records, yielding a large result table. A way to reduce this complexity may be to select from the matching records the one with the largest overlap with the target geometry. An example of this is shown (visually) in Figure 7.4; this is done using `st_join` with argument `largest = TRUE`.

Another way to reduce the result set is to use `aggregate` after a join, to merge all matching records, and union their geometries; see Section 5.4.

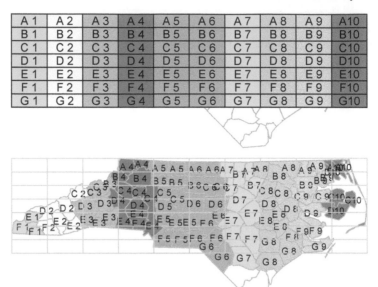

A 1	A 2	A 3	A 4	A 5	A 6	A 7	A 8	A 9	A10
B 1	B 2	B 3	B 4	B 5	B 6	B 7	B 8	B 9	B10
C 1	C 2	C 3	C 4	C 5	C 6	C 7	C 8	C 9	C10
D 1	D 2	D 3	D 4	D 5	D 6	D 7	D 8	D 9	D10
E 1	E 2	E 3	E 4	E 5	E 6	E 7	E 8	E 9	E10
F 1	F 2	F 3	F 4	F 5	F 6	F 7	F 8	F 9	F10
G 1	G 2	G 3	G 4	G 5	G 6	G 7	G 8	G 9	G10

Figure 7.4: Example of `st_join` with `largest = TRUE`: the label of the polygon in the top figure with the largest intersection with polygons in the bottom figure is assigned to the polygons of the bottom figure.

7.2.1   Sampling, gridding, interpolating

Several convenience functions are available in package **sf**, some of which will be discussed here. Function `st_sample` generates a sample of points randomly sampled from target geometries, where target geometries can be point, line, or polygon geometries. Sampling strategies can be (completely) random, regular, or (with polygons) triangular. Chapter 11 explains how spatial sampling (or point pattern simulation) methods available in package **spatstat** are interfaced through `st_sample`.

Function `st_make_grid` creates a square, rectangular, or hexagonal grid over a region, or points with the grid centres or corners. It was used to create the rectangular grid in Figure 7.4.

Function `st_interpolate_aw` "interpolates" area values to new areas, as explained in Section 5.3, both for intensive and extensive variables.

7.3   Ellipsoidal coordinates

Unprojected data have ellipsoidal coordinates, expressed in degrees east and north. As explained in Section 4.1, "straight" lines between points are the curved shortest paths ("geodesic"). By default, **sf** uses geometrical operations from the **s2geometry** library, interfaced through the **s2** package (Dunnington, Pebesma, and Rubak 2023), and we see for instance that the point

```
"POINT(50 50.1)" |> st_as_sfc(crs = "OGC:CRS84") -> pt
```

falls *inside* the polygon:

```
"POLYGON((40 40, 60 40, 60 50, 40 50, 40 40))" |>
  st_as_sfc(crs = "OGC:CRS84") -> pol
st_intersects(pt, pol)
# Sparse geometry binary predicate list of length 1,
# where the predicate was `intersects'
#  1: 1
```

as illustrated by Figure 7.5 (left: straight lines on an orthographic projection centred at the plot area).

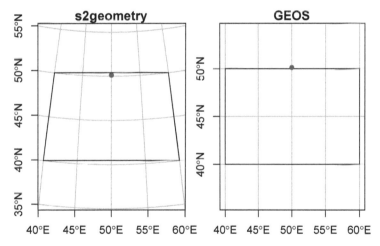

Figure 7.5: Intersection depends on whether we use geodesics/great circle arcs (left: s2) or Cartesian coordinates (right)

If one wants **sf** to use ellipsoidal coordinates as if they are Cartesian coordinates, the use of **s2** can be switched off:

```
old <- sf_use_s2(FALSE)
# Spherical geometry (s2) switched off
st_intersects(pol, pt)
# although coordinates are longitude/latitude, st_intersects
# assumes that they are planar
# Sparse geometry binary predicate list of length 1,
# where the predicate was `intersects'
#  1: (empty)
```

```
sf_use_s2(old) # restore
# Spherical geometry (s2) switched on
```

and the intersection is empty, as illustrated by Figure 7.5 (right: straight lines on an equidistant cylindrical projection). The warning indicates the planar assumption for ellipsoidal coordinates.

Use of **s2** can be switched off for reasons of performance or compatibility with legacy implementations. The underlying libraries, **GEOS** for Cartesian and **s2geometry** for spherical geometry (Figure 1.7) were developed with different motivations, and their use through **sf** differs in some respects:

- certain operations may be much faster in one compared to the other,
- certain functions may be available in only one of them (like `st_relate` being only present in GEOS),
- when using transformers, **GEOS** returns exterior polygon rings noded clockwise (`st_sfc(..., check_ring_dir = TRUE)` can be used to revert to counterclockwise), **s2geometry** retuns them counter-clockwise

7.4   Package stars

Although package **sp** has always had limited support for raster data, over the last decade R package **raster** (Hijmans 2023a) has clearly been dominant as the prime package for powerful, flexible, and scalable raster analysis. The raster data model of package **raster** (and its successor, **terra** (Hijmans 2023b)) is that of a 2D regular raster, or a set of raster layers ("raster stack"). This aligns with the classical static "GIS view", where the world is modelled as a set of layers, each representing a different theme. A lot of data available today however is dynamic, and comes as time series of rasters or raster stacks. A raster stack does not meaningfully reflect this, requiring the user to keep a register of which layer represents what.

Also, the **raster** package and its successor **terra** do an excellent job in scaling computations up to data sizes no larger than the local storage (the computer's hard drives), and doing this fast. Recent datasets, however, including satellite imagery, climate model or weather forecasting data, often no longer fit in local storage (Chapter 9). Package **spacetime** (Pebesma 2012, 2022c) did address the analysis of time series of vector geometries or raster grid cells, but did not extend to higher-dimensional arrays or datasets too large to fit in main memory.

Here, we introduce package **stars** for analysing raster and vector data cubes. The package:

- allows for representing dynamic (time varying) raster stacks
- aims at being scalable, also beyond local disk size
- provides a strong integration of raster functions in the GDAL library
- handles, in addition to regular grids, rotated, sheared, rectilinear, and curvilinear rasters (Figure 1.6)

- provides a tight integration with package **sf**
- handles array data with non-raster spatial dimensions, the *vector data cubes*
- follows the tidyverse design principles

Vector data cubes include for instance time series for simple features, or spatial graph data such as potentially dynamic origin-destination matrices. The concept of spatial vector and raster data cubes was explained in Chapter 6. Irregular spacetime observations can be represented by **sftime** objects provided by package **sftime** (Teickner, Pebesma, and Graeler 2022), which extend **sf** objects with a time column (Section 13.3).

7.4.1 Reading and writing raster data

Raster data typically are read from a file. We use a dataset containing a section of Landsat 7 scene, with the six 30~m-resolution bands (bands 1-5 and 7) for a region covering the city of Olinda, Brazil. We can read the example GeoTIFF file holding a regular, non-rotated grid from the package **stars**:

```
tif <- system.file("tif/L7_ETMs.tif", package = "stars")
library(stars)
# Loading required package: abind
(r <- read_stars(tif))
# stars object with 3 dimensions and 1 attribute
# attribute(s):
#               Min. 1st Qu. Median Mean 3rd Qu. Max.
# L7_ETMs.tif    1     54      69  68.9    86    255
# dimension(s):
#        from   to  offset delta              refsys point x/y
# x      1 349   288776  28.5 SIRGAS 2000 / ... FALSE [x]
# y      1 352  9120761 -28.5 SIRGAS 2000 / ... FALSE [y]
# band   1   6      NA    NA                    NA    NA
```

where we see the offset, cell size, coordinate reference system, and dimensions. The dimension table contains the following fields for each dimension:

- **from**: starting index
- **to**: ending index
- **offset**: dimension value at the start (edge) of the first pixel
- **delta**: cell size; negative **delta** values indicate that pixel index increases with decreasing dimension values
- **refsys**: reference system
- **point**: logical, indicates whether cell values have point support or cell support
- **x/y**: indicates whether a dimension is associated with a spatial raster x- or y-axis

One further field, **values**, is hidden here as it is not used. For regular, rotated, or sheared grids or other regularly discretised dimensions (such as time), **offset** and **delta** are not NA; for irregular cases, **offset** and **delta** are NA and the **values** property contains one of:

- in case of a rectilinear spatial raster or irregular time dimension, the sequence of values or intervals
- in case of a vector data cube, geometries associated with the spatial dimension
- in case of a curvilinear raster, the matrix with coordinate values for each raster cell
- in case of a discrete dimension, the band names or labels associated with the dimension values

The object r is of class **stars** and is a simple list of length one, holding a three-dimensional array:

```
length(r)
# [1] 1
class(r[[1]])
# [1] "array"
dim(r[[1]])
#     x     y band
#   349   352     6
```

and in addition holds an attribute with a dimensions table with all the metadata required to know what the array dimensions refer to, obtained by

```
st_dimensions(r)
```

We can get the spatial extent of the array by

```
st_bbox(r)
#     xmin      ymin      xmax      ymax
#   288776   9110729   298723   9120761
```

Raster data can be written to the local disk using **write_stars**:

```
tf <- tempfile(fileext = ".tif")
write_stars(r, tf)
```

where again the data format (in this case, GeoTIFF) is derived from the file extension. As for simple features, reading and writing uses the GDAL library; the list of available drivers for raster data is obtained by

```
st_drivers("raster")
```

7.4.2   Subsetting **stars** data cubes

Data cubes can be subsetted using the [operator, or using tidyverse verbs. The first option uses [with the following comma-separated arguments:

- attributes first (by name, index, or logical vector)

- followed by each dimension.

This means that r[1:2, 101:200,, 5:10] selects from r attributes 1-2, index 101-200
for dimension 1, and index 5-10 for dimension 3; omitting dimension 2 means that
no subsetting takes place. For attributes, attribute name, index or logical vectors can
be used. For dimensions, logical vectors are not supported. Selecting discontinuous
ranges is supported only when it is a regular sequence. By default, **drop** is FALSE,
when set to **TRUE** dimensions with a single value are dropped:

```
r[,1:100, seq(1, 250, 5), 4] |> dim()
#    x    y band
#  100   50    1
r[,1:100, seq(1, 250, 5), 4, drop = TRUE] |> dim()
#    x    y
#  100   50
```

For selecting particular ranges of dimension *values*, one can use **filter** (after loading
dplyr):

```
library(dplyr, warn.conflicts = FALSE)
filter(r, x > 289000, x < 290000)
# stars object with 3 dimensions and 1 attribute
# attribute(s):
#                 Min. 1st Qu. Median Mean 3rd Qu. Max.
# L7_ETMs.tif        5      51     63 64.3      75  242
# dimension(s):
#          from  to  offset delta             refsys point x/y
# x           1  35  289004  28.5 SIRGAS 2000 / ... FALSE [x]
# y           1 352 9120761 -28.5 SIRGAS 2000 / ... FALSE [y]
# band        1   6       1     1                 NA    NA
```

which changes the offset of the x dimension. Particular cube slices can also be
obtained with **slice**, as in

```
slice(r, band, 3)
# stars object with 2 dimensions and 1 attribute
# attribute(s):
#                 Min. 1st Qu. Median Mean 3rd Qu. Max.
# L7_ETMs.tif       21      49     63 64.4      77  255
# dimension(s):
#    from  to  offset delta             refsys point x/y
# x     1 349  288776  28.5 SIRGAS 2000 / ... FALSE [x]
# y     1 352 9120761 -28.5 SIRGAS 2000 / ... FALSE [y]
```

which drops the singular dimension **band**. **mutate** can be used on **stars** objects to
add new arrays as functions of existing ones, **transmute** drops existing ones.

7.4.3   Cropping

Further subsetting can be done using spatial objects of class **sf**, **sfc** or **bbox**, for instance

```
b <- st_bbox(r) |>
    st_as_sfc() |>
    st_centroid() |>
    st_buffer(units::set_units(500, m))
r[b]
# stars object with 3 dimensions and 1 attribute
# attribute(s):
#                Min. 1st Qu. Median Mean 3rd Qu. Max. NA's
# L7_ETMs.tif      22      54     66 67.7    78.2  174 2184
# dimension(s):
#       from  to  offset delta            refsys point x/y
# x      157 193  288776  28.5 SIRGAS 2000 / ... FALSE [x]
# y      159 194 9120761 -28.5 SIRGAS 2000 / ... FALSE [y]
# band     1   6      NA    NA                  NA    NA
```

selects the circular centre region with a diameter of 500 metre, which for the first band is shown in Figure 7.6,

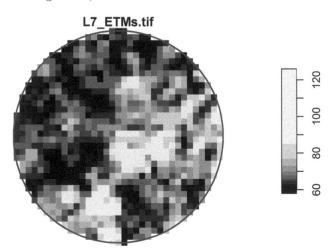

Figure 7.6: Circular centre region of the Landsat 7 scene (band 1)

where we see that pixels outside the spatial object are assigned **NA** values. This object still has dimension indexes relative to the **offset** and **delta** values of **r**; we can reset these to a new **offset** with

```
r[b] |> st_normalize() |> st_dimensions()
#       from to  offset delta            refsys point x/y
# x        1 37  293222  28.5 SIRGAS 2000 / ... FALSE [x]
```

```
# y        1 36 9116258 -28.5 SIRGAS 2000 / ... FALSE [y]
# band     1  6      NA    NA              NA    NA
```

By default, the resulting raster is cropped to the extent of the selection object; an object with the same dimensions as the input object is obtained with

```
r[b, crop = FALSE]
# stars object with 3 dimensions and 1 attribute
# attribute(s):
#              Min. 1st Qu. Median Mean 3rd Qu. Max.   NA's
# L7_ETMs.tif    22      54     66 67.7    78.2  174 731280
# dimension(s):
#      from  to  offset delta            refsys point x/y
# x       1 349  288776  28.5 SIRGAS 2000 / ... FALSE [x]
# y       1 352 9120761 -28.5 SIRGAS 2000 / ... FALSE [y]
# band    1   6      NA    NA              NA    NA
```

Cropping a **stars** object can alternatively be done directly with **st_crop**, as in

```
st_crop(r, b)
```

7.4.4 Redimensioning and combining **stars** objects

Package **stars** uses package **abind** (Plate and Heiberger 2016) for a number of array manipulations. One of them is **aperm** which transposes an array by permuting it. A method for **stars** objects is provided, and

```
aperm(r, c(3, 1, 2))
# stars object with 3 dimensions and 1 attribute
# attribute(s):
#              Min. 1st Qu. Median Mean 3rd Qu. Max.
# L7_ETMs.tif     1      54     69 68.9      86  255
# dimension(s):
#      from  to  offset delta            refsys point x/y
# band    1   6      NA    NA              NA    NA
# x       1 349  288776  28.5 SIRGAS 2000 / ... FALSE [x]
# y       1 352 9120761 -28.5 SIRGAS 2000 / ... FALSE [y]
```

permutes the order of dimensions of the resulting object.

Attributes and dimensions can be swapped, using **split** and **merge**:

```
(rs <- split(r))
# stars object with 2 dimensions and 6 attributes
# attribute(s):
```

```
#        Min. 1st Qu. Median Mean 3rd Qu. Max.
# X1     47      67      78 79.1      89  255
# X2     32      55      66 67.6      79  255
# X3     21      49      63 64.4      77  255
# X4      9      52      63 59.2      75  255
# X5      1      63      89 83.2     112  255
# X6      1      32      60 60.0      88  255
# dimension(s):
#   from   to  offset delta          refsys point x/y
# x    1 349  288776  28.5 SIRGAS 2000 / ... FALSE [x]
# y    1 352 9120761 -28.5 SIRGAS 2000 / ... FALSE [y]
merge(rs, name = "band") |> setNames("L7_ETMs")
# stars object with 3 dimensions and 1 attribute
# attribute(s):
#            Min. 1st Qu. Median Mean 3rd Qu. Max.
# L7_ETMs       1      54     69 68.9      86  255
# dimension(s):
#        from   to  offset delta          refsys point     values
# x         1 349  288776  28.5 SIRGAS 2000 / ... FALSE       NULL
# y         1 352 9120761 -28.5 SIRGAS 2000 / ... FALSE       NULL
# band      1    6      NA    NA                   NA     NA X1,...,X6
#        x/y
# x      [x]
# y      [y]
# band
```

split distributes the **band** dimension over six attributes of a two-dimensional array, **merge** reverses this operation. **st_redimension** can be used for more generic operations, such as splitting a single array dimension over two new dimensions:

```
st_redimension(r, c(x = 349, y = 352, b1 = 3, b2 = 2))
# stars object with 4 dimensions and 1 attribute
# attribute(s):
#                 Min. 1st Qu. Median Mean 3rd Qu. Max.
# L7_ETMs.tif        1      54     69 68.9      86  255
# dimension(s):
#    from   to  offset delta          refsys point
# x     1 349  288776  28.5 SIRGAS 2000 / ... FALSE
# y     1 352 9120761 -28.5 SIRGAS 2000 / ... FALSE
# b1    1    3      NA    NA                   NA    NA
# b2    1    2      NA    NA                   NA    NA
```

Multiple **stars** object with identical dimensions can be combined using **c**. By default, the arrays are combined as additional attributes, but by specifying an **along** argument, the arrays are merged along a new dimension:

```
c(r, r, along = "new_dim")
# stars object with 4 dimensions and 1 attribute
# attribute(s), summary of first 1e+05 cells:
#              Min. 1st Qu. Median Mean 3rd Qu. Max.
# L7_ETMs.tif   47      65      76 77.3      87  255
# dimension(s):
#         from  to  offset delta        refsys point x/y
# x          1 349  288776  28.5 SIRGAS 2000 / ... FALSE [x]
# y          1 352 9120761 -28.5 SIRGAS 2000 / ... FALSE [y]
# band       1   6      NA    NA                NA    NA
# new_dim    1   2      NA    NA                NA    NA
```

the use of this is illustrated in Section 7.5.2.

7.4.5 Extracting point samples, aggregating

A very common use case for raster data cube analysis is the extraction of values at certain locations, or computing aggregations over certain geometries. `st_extract` extracts point values. We will do this for a few randomly sampled points over the bounding box of **r**:

```
set.seed(115517)
pts <- st_bbox(r) |> st_as_sfc() |> st_sample(20)
(e <- st_extract(r, pts))
# stars object with 2 dimensions and 1 attribute
# attribute(s):
#              Min. 1st Qu. Median Mean 3rd Qu. Max.
# L7_ETMs.tif   12    41.8     63   61    80.5  145
# dimension(s):
#           from to            refsys point
# geometry     1 20 SIRGAS 2000 / ...  TRUE
# band         1  6                NA    NA
#                                          values
# geometry POINT (293002 ...,...,POINT (290941 ...
# band                                       NULL
```

which results in a vector data cube with 20 points and 6 bands. (Setting the seed guarantees an identical sample when reproducing, it should not be set generate further randomly generated points.)

Another way of extracting information from data cubes is by aggregating it. One way of doing this is by spatially aggregating values over spatial polygons or lines (Section 6.4). We can for instance compute the maximum pixel value for each of the six bands and for each of the three circles shown in Figure 1.4 (d) by

```
circles <- st_sample(st_as_sfc(st_bbox(r)), 3) |>
    st_buffer(500)
aggregate(r, circles, max)
# stars object with 2 dimensions and 1 attribute
# attribute(s):
#                 Min. 1st Qu. Median Mean 3rd Qu. Max.
# L7_ETMs.tif      73    94.2    117  121     142  205
# dimension(s):
#           from to              refsys point
# geometry     1  3 SIRGAS 2000 / ... FALSE
# band         1  6                 NA    NA
#                                         values
# geometry POLYGON ((2913...,...,POLYGON ((2921...
# band                                       NULL
```

which gives a (vector) data cube with three geometries and six bands. Aggregation over a temporal dimension is done by passing a time variable as the second argument to **aggregate**, as a

- set of time stamps indicating the start of time intervals,
- set of time intervals defined by **make_intervals**, or
- time period like **"weeks"**, **"5 days"**, or **"years"**.

7.4.6   Predictive models

The typical model prediction workflow in R is as follows:

- use a **data.frame** with response and predictor variables (covariates)
- create a model object based on this **data.frame**
- call **predict** with this model object and the **data.frame** with target predictor variable values

Package **stars** provides a **predict** method for **stars** objects that essentially wraps the last step, by creating the **data.frame**, calling the **predict** method for that, and reconstructing a **stars** object with the predicted values.

We will illustrate this with a trivial two-class example mapping land from sea in the example Landsat dataset, using the sample points extracted above, shown in Figure 7.7.

From this figure, we read "by eye" that the points 8, 14, 15, 18, and 19 are on water, the others on land. Using a linear discriminant ("maximum likelihood") classifier, we find model predictions as shown in Figure 7.8 by

```
rs <- split(r)
trn <- st_extract(rs, pts)
trn$cls <- rep("land", 20)
trn$cls[c(8, 14, 15, 18, 19)] <- "water"
```

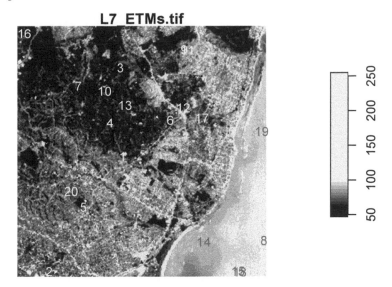

Figure 7.7: Randomly chosen sample locations for training data; red: water, yellow: land

```
model <- MASS::lda(cls ~ ., st_drop_geometry(trn))
pr <- predict(rs, model)
```

Here, we used the `MASS::` prefix to avoid loading **MASS**, as that would mask `select` from **dplyr**. The `split` step is needed to convert the band dimension into attributes, so that they are offered as a set of predictors.

We also see that the layer plotted in Figure 7.8 is a `factor` variable, with class labels.

7.4.7   Plotting raster data

We can use the base `plot` method for **stars** objects, where the plot created with `plot(r)` is shown in Figure 7.9. The default colour scale uses grey tones and stretches them such that colour breaks correspond to data quantiles over all bands ("histogram equalization"). Setting `breaks = "equal"` gives equal width colour breaks; alternatively a sequence of numeric break values can be given. A more familiar view may be the RGB or false colour composite shown in Figure 7.10.

Further details and options are given in Chapter 8.

7.4.8   Analysing raster data

Element-wise mathematical functions (Section 6.3.2) on **stars** objects are just passed on to the arrays. This means that we can call functions and create expressions:

prediction.class

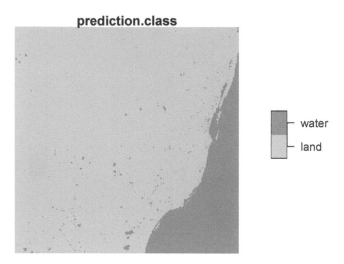

Figure 7.8: Linear discriminant classifier for land/water, based on training data of Figure 7.7

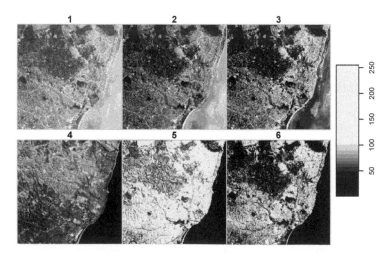

Figure 7.9: Six 30 m Landsat bands downsampled to 90m for Olinda, Br.

RGB   **False colour (NIR-R-G)**

Figure 7.10: Two colour composites

```
log(r)
# stars object with 3 dimensions and 1 attribute
# attribute(s):
#                Min. 1st Qu. Median Mean 3rd Qu. Max.
# L7_ETMs.tif       0    3.99   4.23 4.12    4.45 5.54
# dimension(s):
#      from  to  offset delta            refsys point x/y
# x       1 349  288776  28.5 SIRGAS 2000 / ... FALSE [x]
# y       1 352 9120761 -28.5 SIRGAS 2000 / ... FALSE [y]
# band    1   6      NA    NA                   NA    NA
r + 2 * log(r)
# stars object with 3 dimensions and 1 attribute
# attribute(s):
#                Min. 1st Qu. Median Mean 3rd Qu. Max.
# L7_ETMs.tif       1      62   77.5 77.1    94.9 266
# dimension(s):
#      from  to  offset delta            refsys point x/y
# x       1 349  288776  28.5 SIRGAS 2000 / ... FALSE [x]
# y       1 352 9120761 -28.5 SIRGAS 2000 / ... FALSE [y]
# band    1   6      NA    NA                   NA    NA
```

or even mask out certain values:

```
r2 <- r
r2[r < 50] <- NA
r2
# stars object with 3 dimensions and 1 attribute
# attribute(s):
#                Min. 1st Qu. Median Mean 3rd Qu. Max.   NA's
```

```
# L7_ETMs.tif      50        64       75    79         90  255 149170
# dimension(s):
#        from   to  offset delta                refsys point x/y
# x         1  349  288776   28.5 SIRGAS 2000 / ... FALSE [x]
# y         1  352 9120761  -28.5 SIRGAS 2000 / ... FALSE [y]
# band      1    6      NA     NA                  NA    NA
```

or unmask areas:

```
r2[is.na(r2)] <- 0
r2
# stars object with 3 dimensions and 1 attribute
# attribute(s):
#               Min. 1st Qu. Median Mean 3rd Qu. Max.
# L7_ETMs.tif      0      54     69   63      86  255
# dimension(s):
#        from   to  offset delta                refsys point x/y
# x         1  349  288776   28.5 SIRGAS 2000 / ... FALSE [x]
# y         1  352 9120761  -28.5 SIRGAS 2000 / ... FALSE [y]
# band      1    6      NA     NA                  NA    NA
```

Dimension-wise, we can apply functions to selected array dimensions (Section 6.3.3) of stars objects similar to how **apply** does this to arrays. For instance, we can compute for each pixel the mean of the six band values by

```
st_apply(r, c("x", "y"), mean)
# stars object with 2 dimensions and 1 attribute
# attribute(s):
#         Min. 1st Qu. Median Mean 3rd Qu. Max.
# mean    25.5    53.3   68.3 68.9      82  255
# dimension(s):
#    from   to  offset delta                refsys point x/y
# x     1  349  288776   28.5 SIRGAS 2000 / ... FALSE [x]
# y     1  352 9120761  -28.5 SIRGAS 2000 / ... FALSE [y]
```

A more meaningful function would for instance compute the NDVI (normalised differenced vegetation index):

```
ndvi <- function(b1, b2, b3, b4, b5, b6) (b4 - b3)/(b4 + b3)
st_apply(r, c("x", "y"), ndvi)
# stars object with 2 dimensions and 1 attribute
# attribute(s):
#          Min. 1st Qu.  Median    Mean 3rd Qu.  Max.
# ndvi   -0.753  -0.203 -0.0687 -0.0643   0.187 0.587
# dimension(s):
#    from   to  offset delta                refsys point x/y
```

```
# x      1 349  288776   28.5 SIRGAS 2000 / ... FALSE [x]
# y      1 352 9120761  -28.5 SIRGAS 2000 / ... FALSE [y]
```

Alternatively, one could have defined

```
ndvi2 <- function(x) (x[4]-x[3])/(x[4]+x[3])
```

which is more convenient if the number of bands is large, but also much slower than **ndvi** as it needs to be called for every pixel whereas **ndvi** can be called once for all pixels, or for large chunks of pixels. The mean for each band over the whole image is computed by

```
st_apply(r, c("band"), mean) |> as.data.frame()
#     band mean
# 1      1 79.1
# 2      2 67.6
# 3      3 64.4
# 4      4 59.2
# 5      5 83.2
# 6      6 60.0
```

the result of which is small enough to be printed here as a **data.frame**. In these two examples, entire dimensions disappear. Sometimes, this does not happen (Section 6.3.2); we can for instance compute the three quartiles for each band

```
st_apply(r, c("band"), quantile, c(.25, .5, .75))
# stars object with 2 dimensions and 1 attribute
# attribute(s):
#                Min. 1st Qu. Median Mean 3rd Qu. Max.
# L7_ETMs.tif     32    60.8   66.5 69.8    78.8  112
# dimension(s):
#            from to        values
# quantile     1  3 25%, 50%, 75%
# band         1  6          NULL
```

and see that this *creates* a new dimension, **quantile**, with three values. Alternatively, the three quantiles over the six bands for each pixel are obtained by

```
st_apply(r, c("x", "y"), quantile, c(.25, .5, .75))
# stars object with 3 dimensions and 1 attribute
# attribute(s):
#                Min. 1st Qu. Median Mean 3rd Qu. Max.
# L7_ETMs.tif      4     55   69.2 67.2    81.2  255
# dimension(s):
#           from  to  offset delta            refsys point
```

```
# quantile   1   3      NA     NA                    NA    NA
# x             1 349  288776  28.5 SIRGAS 2000 / ... FALSE
# y             1 352 9120761 -28.5 SIRGAS 2000 / ... FALSE
#                   values x/y
# quantile 25%, 50%, 75%
# x                  NULL [x]
# y                  NULL [y]
```

7.4.9   Curvilinear rasters

There are several reasons why non-regular rasters occur (Figure 1.6). For one, when the data is Earth-bound, a regular raster does not fit the Earth's surface, which is curved. Other reasons are:

- when we convert or transform a regular raster data into another coordinate reference system, it will become curvilinear unless we resample (warp; Section 7.8); warping always comes at the cost of some loss of data and is not reversible
- observation may lead to irregular rasters: for lower-level satellite products, we may have a regular raster in the direction of the satellite (not aligned with x or y), and rectilinear in the direction perpendicular to that (e.g., when the sensor discretises the viewing *angle* in equal parts)

7.4.10   GDAL utils

The GDAL library typically ships with a number of executable binaries, the GDAL command line utilities for data translation and processing. Several of these utilities (all except for those written in Python) are also available as C functions in the GDAL library, through the "GDAL Algorithms C API". If an R package like **sf** that links to the GDAL library uses these C API algorithms, it means that the user no longer needs to install any GDAL binary command line utilities in addition to the R package.

Package **sf** allows calling these C API algorithms through function `gdal_utils`, where the first argument is the name of the utility (stripped from the **gdal** prefix):

- `info` prints information on GDAL (raster) datasets
- `warp` warps a raster to a new raster, possibly in another CRS
- `rasterize` rasterises a vector dataset
- `translate` translates a raster file to another format
- `vectortranslate` (for ogr2ogr) translates a vector file to another format
- `buildvrt` creates a virtual raster tile (a raster created from several files)
- `demprocessing` does various processing steps of digital elevation models (dems)
- `nearblack` converts nearly black/white borders to black
- `grid` creates a regular grid from scattered data
- `mdiminfo` prints information on a multi-dimensional array
- `mdimtranslate` translates a multi-dimensional array into another format

These utilities work on files, and not directly on **sf** or **stars** objects. However, **stars_proxy** objects are essentially pointers to files, and other objects can be written to file. Several of these utilities are (always or optionally) used, for example by **st_mosaic**, **st_warp**, or **st_write**. R package **gdalUtilities** (O'Brien 2022) provides further wrapper functions around **sf::gdal_utils** with function argument names matching the command line arguments of the binary utils.

7.5 Vector data cube examples

7.5.1 Example: aggregating air quality time series

We use a small excert from the European air quality data base to illustrate aggregation operations on vector data cubes. The same data source was used by Gräler, Pebesma, and Heuvelink (2016), and will be used in Chapter 12 and Chapter 13. Daily average PM_{10} values were computed for rural background stations in Germany, over the period 1998-2009.

We can create a **stars** object from the **air** matrix, the **dates** Date vector and the **stations SpatialPoints** objects by

```
load("data/air.rda") # this loads several datasets in .GlobalEnv
dim(air)
# space  time
#    70  4383
stations |>
    st_as_sf(coords = c("longitude", "latitude"), crs = 4326) |>
    st_geometry() -> st
d <- st_dimensions(station = st, time = dates)
(aq <- st_as_stars(list(PM10 = air), dimensions = d))
# stars object with 2 dimensions and 1 attribute
# attribute(s):
#        Min. 1st Qu. Median Mean 3rd Qu. Max.    NA's
# PM10     0    9.92   14.8 17.7      22  274  157659
# dimension(s):
#           from    to    offset  delta refsys point
# station      1    70        NA     NA WGS 84  TRUE
# time         1  4383 1998-01-01 1 days   Date FALSE
#                                            values
# station POINT (9.59 53.7),...,POINT (9.45 49.2)
# time                                         NULL
```

We can see from Figure 7.11 that the time series are quite long, but also have large missing value gaps. Figure 7.12 shows the spatial distribution of measurement stations along with mean PM_{10} values.

We can aggregate these station time series to area means, mostly as a simple exercise. For this, we use the **aggregate** method for **stars** objects

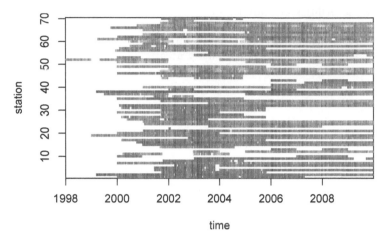

Figure 7.11: Space time diagram of PM_{10} measurements by time and station

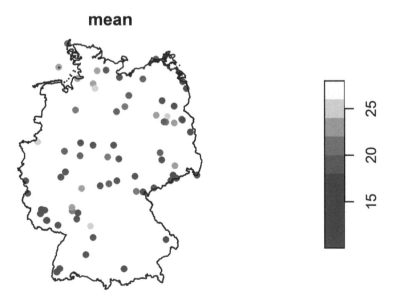

Figure 7.12: Locations of PM_{10} measurement stations, showing mean values

```
(a <- aggregate(aq, de_nuts1, mean, na.rm = TRUE))
# stars object with 2 dimensions and 1 attribute
# attribute(s):
#        Min. 1st Qu. Median Mean 3rd Qu. Max.  NA's
# PM10  1.08    10.9   15.3 17.9    21.8  172 25679
# dimension(s):
#      from   to   offset  delta refsys point
# geom    1   16       NA     NA WGS 84 FALSE
# time    1 4383 1998-01-01 1 days   Date FALSE
#                                          values
# geom MULTIPOLYGON (...,...,MULTIPOLYGON (...
# time                                       NULL
```

and we can now show the maps for six arbitrarily chosen days (Figure 7.13), using

```
library(tidyverse)
a |> filter(time >= "2008-01-01", time < "2008-01-07") |>
    plot(key.pos = 4)
```

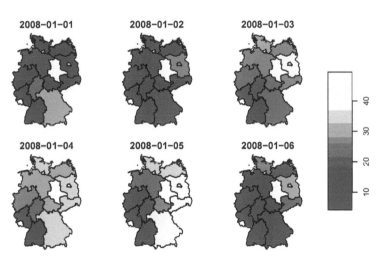

Figure 7.13: Areal mean PM_{10} values, for six days

or create a time series plot of mean values for a single state (Figure 7.14) by

```
library(xts) |> suppressPackageStartupMessages()
plot(as.xts(a)[,4], main = de_nuts1$NAME_1[4])
```

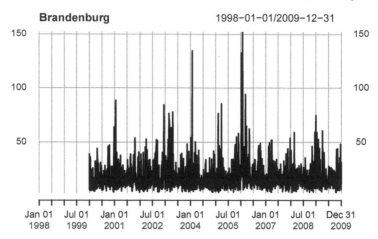

Figure 7.14: Areal mean PM$_{10}$ time series for a single state

7.5.2   Example: Bristol origin-destination data cube

The data used for this example come from Lovelace, Nowosad, and Muenchow (2019), and concern origin-destination (OD) counts: the number of persons going from zone A to zone B, by transportation mode. We have feature geometries in **sf** object **bristol_zones** for the 102 origin and destination regions, shown in Figure 7.15.

OD counts come in a table **bristol_od** with non-zero OD pairs as records, and transportation mode as variables:

```
head(bristol_od)
# # A tibble: 6 x 7
#   o          d            all bicycle   foot car_driver train
#   <chr>      <chr>      <dbl>   <dbl>  <dbl>      <dbl> <dbl>
# 1 E02002985  E02002985    209       5    127         59     0
# 2 E02002985  E02002987    121       7     35         62     0
# 3 E02002985  E02003036     32       2      1         10     1
# 4 E02002985  E02003043    141       1      2         56    17
# 5 E02002985  E02003049     56       2      4         36     0
# 6 E02002985  E02003054     42       4      0         21     0
```

The number of zero OD values is found by subtracting the number of non-zero records from the total number of OD combinations:

```
nrow(bristol_zones)^2 - nrow(bristol_od)
# [1] 7494
```

We will form a three-dimensional vector data cube with origin, destination, and transportation mode as dimensions. For this, we first "tidy" the **bristol_od** table

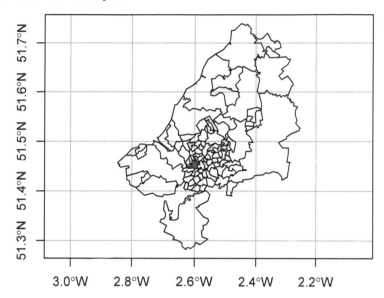

Figure 7.15: Origin destination data zones for Bristol, UK, with zone 33 (E02003043) coloured red

to have origin (o), destination (d), transportation mode (mode), and count (n) as variables, using `pivot_longer`:

```
# create O-D-mode array:
bristol_tidy <- bristol_od |>
    select(-all) |>
    pivot_longer(3:6, names_to = "mode", values_to = "n")
head(bristol_tidy)
# # A tibble: 6 x 4
#   o         d         mode            n
#   <chr>     <chr>     <chr>        <dbl>
# 1 E02002985 E02002985 bicycle          5
# 2 E02002985 E02002985 foot           127
# 3 E02002985 E02002985 car_driver      59
# 4 E02002985 E02002985 train            0
# 5 E02002985 E02002987 bicycle          7
# 6 E02002985 E02002987 foot            35
```

Next, we form the three-dimensional array **a**, filled with zeroes:

```
od <- bristol_tidy |> pull("o") |> unique()
nod <- length(od)
mode <- bristol_tidy |> pull("mode") |> unique()
nmode = length(mode)
```

```
a = array(0L,  c(nod, nod, nmode),
    dimnames = list(o = od, d = od, mode = mode))
dim(a)
# [1] 102 102   4
```

We see that the dimensions are named with the zone names (o, d) and the transportation mode name (mode). Every row of bristol_tidy denotes a single array
entry, and we can use this to fill the non-zero entries of a using the bristol_tidy
table to provide index (o, d and mode) and value (n):

```
a[as.matrix(bristol_tidy[c("o", "d", "mode")])] <-
        bristol_tidy$n
```

To be sure that there is not an order mismatch between the zones in bristol_zones
and the zone names in bristol_tidy, we can get the right set of zones by:

```
order <- match(od, bristol_zones$geo_code)
zones <- st_geometry(bristol_zones)[order]
```

(It happens that the order is already correct, but it is good practice to not assume
this.)

Next, with zones and modes we can create a stars dimensions object:

```
library(stars)
(d <- st_dimensions(o = zones, d = zones, mode = mode))
#       from  to refsys point
# o        1 102 WGS 84 FALSE
# d        1 102 WGS 84 FALSE
# mode     1   4     NA FALSE
#                                               values
# o     MULTIPOLYGON (...,...,MULTIPOLYGON (...
# d     MULTIPOLYGON (...,...,MULTIPOLYGON (...
# mode                        bicycle,...,train
```

and finally build a stars object from a and d:

```
(odm <- st_as_stars(list(N = a), dimensions = d))
# stars object with 3 dimensions and 1 attribute
# attribute(s):
#     Min. 1st Qu. Median Mean 3rd Qu. Max.
# N      0       0      0  4.8       0 1296
# dimension(s):
#       from  to refsys point
# o        1 102 WGS 84 FALSE
# d        1 102 WGS 84 FALSE
```

```
# mode     1    4      NA FALSE
#                                      values
# o    MULTIPOLYGON (...,...,MULTIPOLYGON (...
# d    MULTIPOLYGON (...,...,MULTIPOLYGON (...
# mode                      bicycle,...,train
```

We can take a single slice through this three-dimensional array, for instance for zone 33 (Figure 7.15), by `odm[ , , 33]`, and plot it with

```
plot(adrop(odm[,,33]) + 1, logz = TRUE)
```

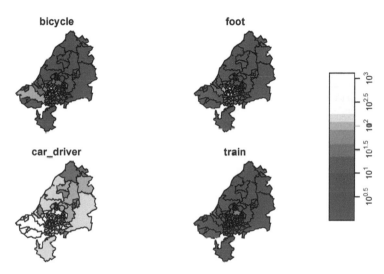

Figure 7.16: OD matrix sliced for destination zone 33, by transportation mode

the result of which is shown in Figure 7.16. Subsetting this way, we take all attributes (there is only one: N) since the first argument is empty, we take all origin regions (second argument empty), we take destination zone 33 (third argument), and all transportation modes (fourth argument empty, or missing).

We plotted this particular zone because it has the largest number of travellers as its destination. We can find this out by summing all origins and travel modes by destination:

```
d <- st_apply(odm, 2, sum)
which.max(d[[1]])
# [1] 33
```

Other aggregations we can carry out include:

Total transportation by OD (102 x 102):

```
st_apply(odm, 1:2, sum)
# stars object with 2 dimensions and 1 attribute
# attribute(s):
#       Min. 1st Qu. Median Mean 3rd Qu. Max.
# sum     0       0      0 19.2      19 1434
# dimension(s):
#   from  to refsys point
# o    1 102 WGS 84 FALSE
# d    1 102 WGS 84 FALSE
#                                        values
# o MULTIPOLYGON (...,...,MULTIPOLYGON (...
# d MULTIPOLYGON (...,...,MULTIPOLYGON (...
```

Origin totals, by mode:

```
st_apply(odm, c(1,3), sum)
# stars object with 2 dimensions and 1 attribute
# attribute(s):
#       Min. 1st Qu. Median Mean 3rd Qu. Max.
# sum     1    57.5    214  490     771 2903
# dimension(s):
#       from  to refsys point
# o        1 102 WGS 84 FALSE
# mode     1   4     NA FALSE
#                                           values
# o       MULTIPOLYGON (...,...,MULTIPOLYGON (...
# mode                         bicycle,...,train
```

Destination totals, by mode:

```
st_apply(odm, c(2,3), sum)
# stars object with 2 dimensions and 1 attribute
# attribute(s):
#       Min. 1st Qu. Median Mean 3rd Qu.  Max.
# sum     0      13    104  490     408 12948
# dimension(s):
#       from  to refsys point
# d        1 102 WGS 84 FALSE
# mode     1   4     NA FALSE
#                                           values
# d       MULTIPOLYGON (...,...,MULTIPOLYGON (...
# mode                         bicycle,...,train
```

Origin totals, summed over modes:

```
o <- st_apply(odm, 1, sum)
```

Destination totals, summed over modes (we had this):

```
d <- st_apply(odm, 2, sum)
```

We plot o and d together after joining them by

```
x <- (c(o, d, along = list(od = c("origin", "destination"))))
plot(x, logz = TRUE)
```

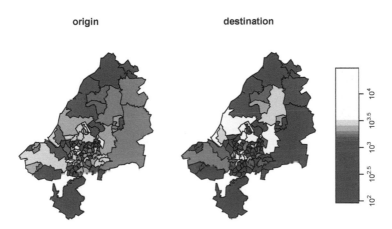

Figure 7.17: Total commutes, summed by origin (left) or destination (right)

the result of which is shown in Figure 7.17.

There is something to say for the argument that such maps give the wrong message, as both amount (colour) and polygon size give an impression of amount. To take out the amount in the count, we can compute densities (count / km^2), by

```
library(units)
a <- set_units(st_area(st_as_sf(o)), km^2)
o$sum_km <- o$sum / a
d$sum_km <- d$sum / a
od <- c(o["sum_km"], d["sum_km"], along =
        list(od = c("origin", "destination")))
plot(od, logz = TRUE)
```

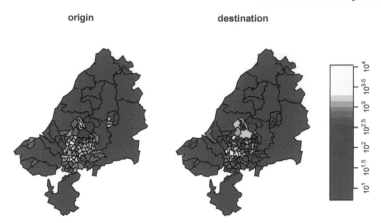

Figure 7.18: Total commutes per square km, by area of origin (left) or destination (right)

shown in Figure 7.18. Another way to normalize these totals would be to divide them by population size.

7.5.3   Tidy array data

The *tidy data* paper (Wickham 2014b) may suggest that such array data should be processed not as an array, but in a long (unnormalised) table form where each row holds (region, class, year, value), and it is always good to be able to do this. For primary handling and storage, however, this is often not an option, because:

- a lot of array data are collected or generated as array data, for instance by imagery or other sensory devices, or by climate models
- it is easier to derive the long table form from the array than vice versa
- the long table form requires much more memory, since the space occupied by dimension values is $O(\Pi n_i)$, rather than $O(\Sigma n_i)$, with n_i the cardinality (size) of dimension i
- when missing-valued cells are dropped, the long table form loses the implicit indexing of the array form

To put this argument to the extreme, consider for instance that all image, video and sound data are stored in array form; few people would make a real case for storing them in a long table form instead. Nevertheless, R packages like **tsibble** (Wang et al. 2022) take this approach, and have to deal with ambiguous ordering of multiple records with identical time steps for different spatial features and index them, which is solved for both *automatically* by using the array form – at the cost of using dense arrays, in package **stars**.

Package **stars** tries to follow the tidy manifesto to handle array sets and has particularly developed support for the case where one or more of the dimensions refer to space and/or time.

7.5.4 File formats for vector data cubes

Regular table forms, including the long table form, are possible but clumsy to use: the origin-destination data example above and Chapter 13 illustrate the complexity of recreating a vector data cube from table forms. Array formats like NetCDF or Zarr are designed for storing array data. They can however be used for *any* data structure, and carry the risk that files once written are hard to reuse. For vector cubes that have a *single* geometry dimension that consists of either points, (multi)linestrings or (multi)polygons, the CF conventions (Eaton et al. 2022) describe a way to encode such geometries. `stars::read_mdim` and `stars::write_mdim` can read and write vector data cubes following these conventions.

7.6 Raster-to-vector, vector-to-raster

Section 1.3 already showed some examples of raster-to-vector and vector-to-raster conversions. This section will add some code details and examples.

7.6.1 Vector-to-raster

`st_as_stars` is meant as a method to transform objects into `stars` objects. However, not all stars objects are `raster` objects, and the method for `sf` objects creates a vector data cube with the geometry as its spatial (vector) dimension, and attributes as attributes. When given a feature *geometry* (`sfc`) object, `st_as_stars` will rasterise it, as shown in Section 7.8 and in Figure 7.19.

```
file <- system.file("gpkg/nc.gpkg", package="sf")
read_sf(file) |>
    st_geometry() |>
    st_as_stars() |>
    plot(key.pos = 4)
```

Here, `st_as_stars` can be parameterised to control cell size, number of cells, and/or extent. The cell values returned are 0 for cells with centre point outside the geometry and 1 for cell with centre point inside or on the border of the geometry. Rasterising existing features is done using `st_rasterize`, as also shown in Figure 1.5:

```
library(dplyr)
read_sf(file) |>
    mutate(name = as.factor(NAME)) |>
    select(SID74, SID79, name) |>
    st_rasterize()
# stars object with 2 dimensions and 3 attributes
# attribute(s):
#       SID74           SID79               name
```

values

Figure 7.19: Rasterising vector geometry using `st_as_stars`

```
# Min.   : 0      Min.   : 0      Sampson :   655
# 1st Qu.: 3      1st Qu.: 3      Columbus:   648
# Median : 5      Median : 6      Robeson :   648
# Mean   : 8      Mean   :10      Bladen  :   604
# 3rd Qu.:10      3rd Qu.:13      Wake    :   590
# Max.   :44      Max.   :57      (Other) :30952
# NA's   :30904   NA's   :30904   NA's    :30904
# dimension(s):
#   from  to   offset    delta refsys point x/y
# x    1 461 -84.3239  0.0192484  NAD27 FALSE [x]
# y    1 141  36.5896 -0.0192484  NAD27 FALSE [y]
```

Similarly, line and point geometries can be rasterised, as shown in Figure 7.20.

```
read_sf(file) |>
    st_cast("MULTILINESTRING") |>
    select(CNTY_ID) |>
    st_rasterize() |>
    plot(key.pos = 4)
```

7.7   Coordinate transformations and conversions

7.7.1   `st_crs`

Spatial objects of class **sf** or **stars** contain a coordinate reference system that can be retrieved or replaced with `st_crs`, or can be set or replaced in a pipe with `st_set_crs`. Coordinate reference systems can be set with an EPSG code, like `st_crs(4326)` which will be converted to `st_crs('EPSG:4326')`, or with a PROJ.4 string like

CNTY_ID

Figure 7.20: Rasterising the North Carolina county boundaries

"+proj=utm +zone=25 +south", a name like "WGS84", or a name preceded by an authority like "OGC:CRS84". Alternatives include a coordinate reference system definition in WKT, WKT-2 (Section 2.5) or PROJJSON. The object returned by st_crs contains two fields:

- wkt with the WKT-2 representation
- input with the user input, if any, or a human readable description of the coordinate reference system, if available

Note that PROJ.4 strings can be used to *define* some coordinate reference systems, but they cannot be used to *represent* coordinate reference systems. Conversion of a WKT-2 in a crs object to a proj4string using the $proj4string method, as in

```
x <- st_crs("OGC:CRS84")
x$proj4string
# [1] "+proj=longlat +datum=WGS84 +no_defs"
```

may succeed but is not in general lossless or invertible. Using PROJ.4 strings, for instance to *define* a parameterised, projected coordinate reference system is fine as long as it is associated with the WGS84 datum.

7.7.2 st_transform, sf_project

Coordinate transformations or conversions (Section 2.4) for sf or stars objects are carried out with st_transform, which takes as its first argument a spatial object of class sf or stars that has a coordinate reference system set, and as a second argument with an crs object (or something that can be converted to it with st_crs). When PROJ finds more than one possibility to transform or convert from the source crs to the target crs, it chooses the one with the highest declared accuracy. More fine-grained control over the options is explained in Section 7.7.5. For stars objects with regular raster dimensions, st_transform will *only* transform coordinates and always result in a curvilinear grid. `st_warp` can be used to create a regular raster in a new coordinate reference system, using regridding (Section 7.8).

A lower-level function to transform or convert coordinates *not* in **sf** or **stars** objects is `sf_project`: it takes a matrix with coordinates and a source and target **crs**, and it returns transformed or converted coordinates.

7.7.3   `sf_proj_info`

Function `sf_proj_info` can be used to query the available projections, ellipsoids, units and prime meridians available in the PROJ software. It takes a single parameter, **type**, which can have the following values:

- **type** = "**proj**" lists the short and long names of available projections; short names can be used in a "+proj=name" string
- **type** = "**ellps**" lists the available ellipses, with name, long name, and ellipsoidal parameters
- **type** = "**units**" lists the available length units, with conversion constant to meters
- **type** = "**prime_meridians**" lists the prime meridians with their position with respect to the Greenwich meridian

7.7.4   Datum grids, proj.db, cdn.proj.org, local cache

Datum grids (Section 2.4) can be installed locally, or be read from the PROJ datum grid CDN at https://cdn.proj.org/. If installed locally, they are read from the PROJ search path, which is shown by

```
sf_proj_search_paths()
# [1] "/home/edzer/.local/share/proj"
# [2] "/usr/share/proj"
```

The main PROJ database is `proj.db`, an sqlite3 database typically found at

```
paste0(tail(sf_proj_search_paths(), 1), .Platform$file.sep,
       "proj.db")
# [1] "/usr/share/proj/proj.db"
```

which can be read. The version of the snapshot of the EPSG database included in each PROJ release is stated in the "**metadata**" table of `proj.db`; the version of the PROJ runtime used by **sf** is shown by

```
sf_extSoftVersion()["PROJ"]
#    PROJ
# "9.1.1"
```

If for a particular coordinate transformation datum grids are not locally found, PROJ will search for online datum grids in the PROJ CDN when

```
sf_proj_network()
# [1] FALSE
```

returns **TRUE**. By default it is set to **FALSE**, but

```
sf_proj_network(TRUE)
# [1] "https://cdn.proj.org"
```

sets it to **TRUE** and returns the URL of the network resource used. This resource can also be set to another resource, that may be faster or less limited.

After querying a datum grid on the CDN, PROJ writes the *portion* of the grid queried (not, by default, the entire grid) to a local cache, which is another sqlite3 database found locally in a user directory that is listed by

```
list.files(sf_proj_search_paths()[1], full.names = TRUE)
# [1] "/home/edzer/.local/share/proj/cache.db"
```

and that will be searched first in subsequent datum grid queries.

7.7.5 Transformation pipelines

Internally, PROJ uses a so-called *coordinate operation pipeline*, to represent the sequence of operations to get from a source CRS to a target CRS. Given multiple options to go from source to target, **st_transform** chooses the one with highest accuracy. We can query the options available by **sf_proj_pipelines**:

```
(p <- sf_proj_pipelines("OGC:CRS84", "EPSG:22525"))
# Candidate coordinate operations found:  5
# Strict containment:     FALSE
# Axis order auth compl:  FALSE
# Source:   OGC:CRS84
# Target:   EPSG:22525
# Best instantiable operation has accuracy: 2 m
# Description: axis order change (2D) + Inverse of Corrego
#              Alegre 1970-72 to WGS 84 (2) + UTM
#              zone 25S
# Definition:  +proj=pipeline +step +proj=unitconvert
#              +xy_in=deg +xy_out=rad +step +inv
#              +proj=hgridshift
#              +grids=br_ibge_CA7072_003.tif +step
#              +proj=utm +zone=25 +south
#              +ellps=intl
```

and see that pipeline with the highest accuracy is summarised; we can see that it specifies use of a datum grid. Had we not switched on the network search, we would have obtained a different result:

```
sf_proj_network(FALSE)
# character(0)
sf_proj_pipelines("OGC:CRS84", "EPSG:22525")
# Candidate coordinate operations found:  5
# Strict containment:     FALSE
# Axis order auth compl:  FALSE
# Source:  OGC:CRS84
# Target:  EPSG:22525
# Best instantiable operation has accuracy: 2 m
# Description: axis order change (2D) + Inverse of Corrego
#              Alegre 1970-72 to WGS 84 (2) + UTM
#              zone 25S
# Definition:  +proj=pipeline +step +proj=unitconvert
#              +xy_in=deg +xy_out=rad +step
#              +proj=push +v_3 +step +proj=cart
#              +ellps=WGS84 +step +proj=helmert
#              +x=206.05 +y=-168.28 +z=3.82 +step
#              +inv +proj=cart +ellps=intl +step
#              +proj=pop +v_3 +step +proj=utm
#              +zone=25 +south +ellps=intl
# Operation 4 is lacking 1 grid with accuracy 2 m
# Missing grid: br_ibge_CA7072_003.tif
# URL: https://cdn.proj.org/br_ibge_CA7072_003.tif
```

and a report that a datum grid is missing. The object returned by **sf_proj_pipelines** is a sub-classed **data.frame**, with columns

```
names(p)
# [1] "id"           "description"   "definition"
# [4] "has_inverse"  "accuracy"      "axis_order"
# [7] "grid_count"   "instantiable"  "containment"
```

and we can list for instance the accuracies by

```
p |> pull(accuracy)
# [1]  2  5  5  8 NA
```

Here, **NA** refers to "ballpark accuracy", which may be anything in the 30-120 m range:

```
p |> filter(is.na(accuracy))
# Candidate coordinate operations found:  1
# Strict containment:     FALSE
```

```
# Axis order auth compl:  FALSE
# Source:  OGC:CRS84
# Target:  EPSG:22525
# Best instantiable operation has only ballpark accuracy
# Description: Ballpark geographic offset from WGS 84 (CRS84)
#              to Corrego Alegre 1970-72 + UTM
#              zone 25S
# Definition:  +proj=pipeline +step +proj=unitconvert
#              +xy_in=deg +xy_out=rad +step
#              +proj=utm +zone=25 +south
#              +ellps=intl
```

The default, most accurate pipeline chosen by **st_transform** can be overridden by specifying **pipeline** argument, as selected from the set of options in **p$definition**.

7.7.6 Axis order and direction

As mentioned in Section 2.5, **EPSG:4326** defines the first axis to be associated with latitude and the second with longitude; this is also the case for a number of other ellipsoidal coordinate reference systems. Although this is how the authority (EPSG) prescribes this, it is not how most datasets are currently stored. As most other software, package **sf** by default ignores this and interprets ellipsoidal coordinate pairs as (longitude, latitude) by default. If however data needs to be read from a data source that is compliant to the authority, for instance from a WFS service, one can set

```
st_axis_order(TRUE)
```

to globally instruct **sf**, when calling GDAL and PROJ routines, that authority compliance (latitude, longitude order) is assumed. It is anticipated that problems may happen in case of authority compliance, for instance when with plotting data. The **plot** method for **sf** objects respects the axis order flag and will swap coordinates using the transformation pipeline "**+proj=pipeline +step +proj=axisswap +order=2,1**" before plotting them, but **geom_sf** in **ggplot2** has not been modified to do this. As mentioned earlier, the axis order ambiguity of **EPSG:4326** is resolved by replacing it with **OGC:CRS84**.

Independent from the axis order, not all coordinate reference systems have an axis direction that is positive in North and East directions. Most plotting functions in R will not work with data that have axes defined in opposite directions. Information on axes directions and units can be retrieved by

```
st_crs(4326)$axes
#                  name orientation convfactor
# 1  Geodetic latitude           1     0.0175
# 2 Geodetic longitude           3     0.0175
st_crs(4326)$ud_unit
```

```
# 1 [°]
st_crs("EPSG:2053")$axes
#         name orientation convfactor
# 1  Westing             4          1
# 2 Southing             2          1
st_crs("EPSG:2053")$ud_unit
# 1 [m]
```

7.8   Transforming and warping rasters

When using **st_transform** on a raster dataset, as in

```
tif <- system.file("tif/L7_ETMs.tif", package = "stars")
read_stars(tif) |>
    st_transform('OGC:CRS84')
# stars object with 3 dimensions and 1 attribute
# attribute(s):
#               Min. 1st Qu. Median Mean 3rd Qu. Max.
# L7_ETMs.tif      1      54     69 68.9      86  255
# dimension(s):
#      from  to refsys point                         values
# x       1 349 WGS 84 FALSE [349x352] -34.9165,...,-34.8261
# y       1 352 WGS 84 FALSE [349x352] -8.0408,...,-7.94995
# band    1   6     NA    NA                           NULL
#      x/y
# x    [x]
# y    [y]
# band
# curvilinear grid
```

we see that a *curvilinear* is created, which means that for every grid cell the coordinates are computed in the new CRS, which no longer form a *regular* grid. Plotting such data is extremely slow, as small polygons are computed for every grid cell and then plotted. The advantage of this is that no information is lost: grid cell values remain identical after the projection.

When we start with a raster on a regular grid and want to obtain a *regular* grid in a new coordinate reference system, we need to *warp* the grid: we need to recreate a grid at new locations and use some rule to assign values to new grid cells. Rules can involve using the nearest value or using some form of interpolation. This operation is not lossless and not invertible.

The best approach for warping is to specify the target grid as a **stars** object. When only a target CRS is specified, default options for the target grid are picked that may be completely inappropriate for the problem at hand. An example workflow that uses only a target CRS is

```
read_stars(tif) |>
    st_warp(crs = st_crs('OGC:CRS84')) |>
    st_dimensions()
#        from  to    offset        delta refsys x/y
# x         1 350 -34.9166  0.000259243 WGS 84 [x]
# y         1 352 -7.94982 -0.000259243 WGS 84 [y]
# band      1   6       NA           NA     NA
```

which creates a pretty close raster, but then the transformation is also relatively modest. For a workflow that creates a target raster first, here with exactly the same number of rows and columns as the original raster, one could use:

```
r <- read_stars(tif)
grd <- st_bbox(r) |>
        st_as_sfc() |>
        st_transform('OGC:CRS84') |>
        st_bbox() |>
        st_as_stars(nx = dim(r)["x"], ny = dim(r)["y"])
st_warp(r, grd)
# stars object with 3 dimensions and 1 attribute
# attribute(s):
#              Min. 1st Qu. Median Mean 3rd Qu. Max. NA's
# L7_ETMs.tif     1      54     69 68.9      86  255 6180
# dimension(s):
#        from  to    offset        delta refsys x/y
# x         1 349 -34.9166  0.000259666 WGS 84 [x]
# y         1 352 -7.94982 -0.000258821 WGS 84 [y]
# band      1   6       NA           NA     NA
```

where we see that grid resolution in x and y directions slightly varies.

7.9 Exercises

Use R to solve the following exercises.

1. Find the names of the **nc** counties that intersect **LINESTRING(-84 35,-78 35)**; use [for this, and as an alternative use **st_join** for this.
2. Repeat this after setting **sf_use_s2(FALSE)**, and *compute* the difference (hint: use **setdiff**), and colour the counties of the difference using colour '#88000088'.
3. Plot the two different lines in a single plot; note that R will plot a straight line always straight in the projection currently used; **st_segmentize** can

be used to add points on a straight line, or on a great circle for ellipsoidal coordinates.

4. NDVI, normalised differenced vegetation index, is computed as (NIR-R)/(NIR+R), with NIR the near infrared and R the red band. Read the L7_ETMs.tif file into object x, and distribute the band dimensions over attributes by split(x, "band"). Then, add attribute NDVI to this object by using an expression that uses the NIR (band 4) and R (band 3) attributes directly.

5. Compute NDVI for the L7_ETMs.tif image by reducing the band dimension, using st_apply and a function ndvi = function(x) { (x[4]-x[3])/(x[4]+x[3]) }. Plot the result, and write the result to a GeoTIFF.

6. Use st_transform to transform the stars object read from L7_ETMs.tif to OGC:CRS84. Print the object. Is this a regular grid? Plot the first band using arguments axes=TRUE and border=NA, and explain why this takes such a long time.

7. Use st_warp to warp the L7_ETMs.tif object to OGC:CRS84, and plot the resulting object with axes=TRUE. Why is the plot created much faster than after st_transform?

8. Using a vector representation of the raster L7_ETMs, plot the intersection with a circular area around POINT(293716 9113692) with radius 75 m, and compute the area-weighted mean pixel values for this circle. Compare the area-weighted values with those obtained by aggregate using the vector data, and by aggregate using the raster data, using exact=FALSE (default) and exact=FALSE. Explain the differences.

8

Plotting spatial data

Together with timelines, maps belong to the most powerful graphs, perhaps because we can immediately relate to where we are, or once have been, on the space of the plot. Two recent books on visualisation (Healy 2018; Wilke 2019) contain chapters on visualising geospatial data or maps. Here, we will not try to point out which maps are good and which are bad, but rather a number of possibilities for creating them, challenges along the way, and possible ways to mitigate them.

8.1 Every plot is a projection

The world is round, but plotting devices are flat. As mentioned in Section 2.2.2, any time we visualise, in any way, the world on a flat device, we project: we convert ellipsoidal coordinates into Cartesian coordinates. This includes the cases where we think we "do nothing" as in Figure 8.1 (left), or where we show the world "as it is", as one would see it from space (Figure 8.1, right).

The left plot of Figure 8.1 was obtained by

```
library(sf)
library(rnaturalearth)
w <- ne_countries(scale = "medium", returnclass = "sf")
plot(st_geometry(w))
```

indicating that this is the default projection for global data with ellipsoidal coordinates:

```
st_is_longlat(w)
# [1] TRUE
```

The projection taken in Figure 8.1 (left) is the equirectangular (or equidistant cylindrical) projection, which maps longitude and latitude linearly to the x- and y-axis, keeping an aspect ratio of 1. If we would do this for smaller areas not on the equator, then it would make sense to choose a plot ratio such that one distance unit E-W equals one distance unit N-S at the centre of the plotted area, and this is the default behaviour of the **plot** method for unprojected **sf** or **stars** datasets, as well as the default for **ggplot2::geom_sf** (Section 8.4).

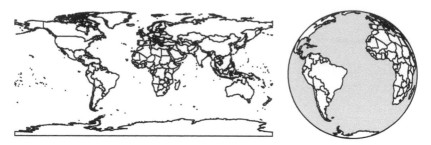

Figure 8.1: Earth country boundaries; left: mapping long/lat linearly to x and y (plate carrée); right: as seen from an infinite distance (orthographic)

We can also carry out this projection before plotting. Say we want to plot Germany, then after loading the (rough) country outline, we use **st_transform** to project:

```
DE <- st_geometry(ne_countries(country = "germany",
                               returnclass = "sf"))
DE |> st_transform("+proj=eqc +lat_ts=51.14 +lon_0=90w") ->
  DE.eqc
```

Here, **eqc** refers to the "equidistant cylindrical" projection of PROJ. The projection parameter here is **lat_ts**, the latitude of *true scale*, where one length unit N-S equals one length unit E-W. This was chosen at the middle of the bounding box latitudes

```
# [1] 51.14
```

We plot both maps in Figure 8.2, and they look identical up to the values along the axes: degrees for ellipsoidal (left) and metres for projected (Cartesian, right) coordinates.

8.1.1 What is a good projection for my data?

There is unfortunately no silver bullet here. Projections that maintain all distances do not exist; only globes do. The most used projections try to preserve:

- areas (equal area)
- directions (conformal, such as *Mercator*)
- some properties of distances (*equirectangular* preserves distances along meridians, *azimuthal equidistant* preserves distances to a central point)

or some compromise of these. Parameters of projections decide what is shown in the centre of a map and what is shown on the fringes, which areas are up and which are down, and which areas are most enlarged. All these choices are in the end political decisions.

It is often entertaining and at times educational to play around with the different projections and understand their consequences. When the primary purpose of the map however is not to entertain or educate projection varieties, it may be preferable

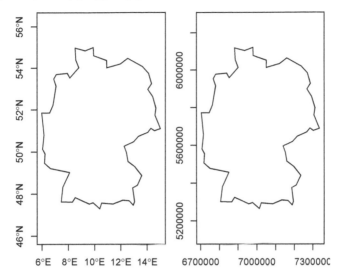

Figure 8.2: Germany in equirectangular projection: with axis units degrees (left) and metres in the equidistant cylindrical projection (right)

to choose a well-known or less surprising projection and move the discussion which projection to use to a decision process of its own. For global maps however, in almost all cases, equal area projections are preferred over plate carrée or web Mercator projections.

8.2   Plotting points, lines, polygons, grid cells

Since maps are just a special form of plots of statistical data, the usual rules hold. Frequently occurring challenges include:

- polygons may be very small, and vanish when plotted
- depending on the data, polygons for different features may well overlap, and be visible only partially; using transparent fill colours may help identify them
- when points are plotted with symbols, they may easily overlap and be hidden; density maps (Chapter 11) may be more helpful
- lines may be hard to read when coloured and may overlap regardless the line width

8.2.1   Colours

When plotting polygons filled with colours, one has the choice to plot polygon boundaries or to suppress these. If polygon boundaries draw too much attention, an alternative is to colour them in a grey tone, or another colour that does not interfere with the fill colours. When suppressing boundaries entirely, polygons with (nearly) identical colours will no longer be visually distinguishable. If the property indicating the fill colour is constant over the region, such as land cover type, then this is not a problem, but if the property is an aggregation then the region over which it was aggregated gets lost, and by that the proper interpretation. Especially for extensive variables, such as the amount of people living in a polygon, this strongly misleads. But even with polygon boundaries, using filled polygons for extensive variables may not be a good idea because the map colours conflate amount and area size.

The use of continuous colour scales that have no noticeable colour breaks for continuously varying variables may look attractive, but is often more fancy than useful:

- it is impracticable to match a colour on the map with a legend value
- colour ramps often stretch non-linearly over the value range, making it hard to convey magnitude

Only for cases where the identification of values is less important than the continuity of the map, such as the colouring of a high resolution digital terrain model, it does serve its goal. Good colours scales and palettes are found in functions **hcl.colors** or **palette.colors**, and in packages **RColorBrewer** (Neuwirth 2022), **viridis** (Garnier 2021), or **colorspace** (Ihaka et al. 2023; Zeileis et al. 2020).

8.2.2   Colour breaks: `classInt`

When plotting continuous geometry attributes using a limited set of colours (or symbols), classes need to be made from the data. R package **classInt** (Bivand 2022a) provides a number of methods to do so. The default method is "quantile":

```
library(classInt)
# set.seed(1) if needed ?
r <- rnorm(100)
(cI <- classIntervals(r))
# style: quantile
#    one of 1.49e+10 possible partitions of this variable into 8 classes
#      [-2.25,-1.08)    [-1.08,-0.681)    [-0.681,-0.169)
#            13               12                13
# [-0.169,-0.0617)    [-0.0617,0.202)     [0.202,0.454)
#            12               12                13
#      [0.454,1.05)       [1.05,2.14]
#            12               13
cI$brks
# [1] -2.2527 -1.0799 -0.6814 -0.1685 -0.0617  0.2018  0.4536
# [8]  1.0537  2.1392
```

it takes argument **n** for the number of intervals, and a **style** that can be one of "fixed", "sd", "equal", "pretty", "quantile", "kmeans", "hclust", "bclust", "fisher" or "jenks". Style "pretty" may not obey n; if n is missing, **nclass.Sturges** is used; two other methods are available for choosing **n** automatically. If the number of observations is greater than 3000, a 10% sample is used to create the breaks for "fisher" and "jenks".

8.2.3 Graticule and other navigation aids

A graticule is a network of lines on a map that follow constant latitude or longitude. Figure 1.1 shows a graticule drawn in grey, on Figure 1.2 it is white. Graticules are often drawn in maps to give place reference. In our first map in Figure 1.1 we can read that the area plotted is near $35°$ North and $80°$ West. Had we plotted the lines in the projected coordinate system, they would have been straight and their actual numbers would not have been very informative, apart from giving an interpretation of size or distances when the unit is known, and familiar to the map reader. Graticules also shed light on which projection was used: equirectangular or Mercator projections have straight vertical and horizontal lines, conic projections have straight but diverging meridians, and equal area projections may have curved meridians.

On Figure 8.1 and most other maps the real navigation aid comes from geographical features like the state outline, country outlines, coast lines, rivers, roads, railways and so on. If these are added sparsely and sufficiently, a graticule can as well be omitted. In such cases, maps look good without axes, tics, and labels, leaving up a lot of plotting space to be filled with actual map data.

8.3 Base plot

The `plot` method for `sf` and `stars` objects try to make quick, useful, exploratory plots; for higher quality plots and more configurability, alternatives with more control and/or better defaults are offered for instance by packages **ggplot2** (Wickham et al. 2022), **tmap** (Tennekes 2022, 2018), or **mapsf** (Giraud 2022).

By default, the plot method tries to plot "all" it is given. This means that:

- given a geometry only (`sfc`), the geometry is plotted, without colours
- given a geometry and an attribute, the geometry is coloured according to the values of the attribute, using a qualitative colour scale for `factor` or `logical` attributes and a continuous scale otherwise, and a colour key is added
- given multiple attributes, multiple maps are plotted, each with a colour scale but a key is by default omitted, as colour assignment is done on a per sub-map basis
- for `stars` objects with multiple attributes, only the first attribute is plotted; for three-dimensional raster cubes, all slices over the third dimension are plotted as sub-plots

8.3.1 Adding to plots with legends

The `plot` methods for `stars` and `sf` objects may show a colour key on one of the sides (Figure 1.1). To do this with `base::plot`, the plot region is split in two and two plots are created: one with the map, and one with the legend. By default, the `plot` function resets the graphics device (using `layout(matrix(1))` so that subsequent plots are not hindered by the device being split in two, but this prevents adding graphic elements subsequently. To *add* to an existing plot with a colour legend, the device reset needs to be prevented by using `reset = FALSE` in the `plot` command, and using `add = TRUE` in subsequent calls to `plot`. An example is

```
library(sf)
nc <- read_sf(system.file("gpkg/nc.gpkg", package = "sf"))
plot(nc["BIR74"], reset = FALSE, key.pos = 4)
plot(st_buffer(nc[1,1], units::set_units(10, km)), col = 'NA',
     border = 'red', lwd = 2, add = TRUE)
```

BIR74

Figure 8.3: Annotating base plots with a legend

which is shown in Figure 8.3. Annotating **stars** plots can be done in the same way when a *single* stars layer is shown. Annotating **stars** facet plots with multiple cube slices can be done by adding a "hook" function that will be called on every slice shown, as in

```
library(stars)
# Loading required package: abind
system.file("tif/L7_ETMs.tif", package = "stars") |>
    read_stars() -> r
st_bbox(r) |> st_as_sfc() |> st_sample(5) |>
    st_buffer(300) -> circ
hook <- function() {
    plot(circ, col = NA, border = 'yellow', add = TRUE)
}
plot(r, hook = hook, key.pos = 4)
# downsample set to 1
```

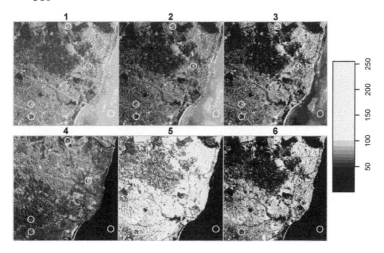

Figure 8.4: Annotated multi-slice stars plot

and as shown in Figure 8.4. Hook functions have access to facet parameters, facet label and bounding box.

Base plot methods have access to the resolution of the screen device, and hence the base plot method for `stars` and `stars_proxy` object will downsample dense rasters and only plot pixels at a density that makes sense for the device available.

8.3.2 Projections in base plots

The base plot method plots data with ellipsoidal coordinates using the equirectangular projection, using a latitude parameter equal to the middle latitude of the data bounding box (Figure 8.2). To control this parameter, either a projection to another equirectangular can be applied before plotting, or the parameter `asp` can be set to override: `asp=1` would lead to plate carrée (Figure 8.1) left. Subsequent plots need to be in the same coordinate reference system in order to make sense with over-plotting; this is not being checked.

8.3.3 Colours and colour breaks

In base plots, argument `nbreaks` can be used to set the number of colour breaks and argument `breaks` either to the numeric vector with actual breaks, or to a style value for the `style` argument in `classInt::classIntervals`.

8.4 Maps with ggplot2

Package **ggplot2** (Wickham et al. 2022; Wickham 2016) can create more complex and nicer looking graphs; it has a geometry `geom_sf` that was developed in conjunction with the development of **sf** and helps creating beautiful maps. An introduction to this is found in Moreno and Basille (2018). A first example is shown in Figure 1.2. The code used for this plot is:

```
library(tidyverse) |> suppressPackageStartupMessages()
nc.32119 <- st_transform(nc, 32119)
year_labels <-
    c("SID74" = "1974 - 1978", "SID79" = "1979 - 1984")
nc.32119 |> select(SID74, SID79) |>
    pivot_longer(starts_with("SID")) -> nc_longer


ggplot() + geom_sf(data = nc_longer, aes(fill = value), linewidth = 0.4) +
  facet_wrap(~ name, ncol = 1,
             labeller = labeller(name = year_labels)) +
  scale_y_continuous(breaks = 34:36) +
  scale_fill_gradientn(colours = sf.colors(20)) +
  theme(panel.grid.major = element_line(colour = "white"))
```

where we see that two attributes had to be stacked (**pivot_longer**) before plotting them as facets: this is the idea behind "tidy" data, and the **pivot_longer** method for **sf** objects automatically stacks the geometry column too.

Because **ggplot2** creates graphics *objects* before plotting them, it can control the coordinate reference system of all elements involved, and will transform or convert all subsequent objects to the coordinate reference system of the first. It will also draw a graticule for the (default) thin white lines on a grey background, and uses a datum (by default: WGS84) for this. **geom_sf** can be combined with other geoms, for instance to allow for annotating plots.

For package **stars**, a **geom_stars** has, at the moment of writing this, rather limited scope: it uses **geom_sf** for map layout and vector data cubes, and adds **geom_raster** for regular rasters and **geom_rect** for rectilinear rasters. It downsamples if the user specifies a downsampling rate, but has no access to the screen dimensions to automatically choose a downsampling rate. This may be just enough, for instance Figure 8.5 is created by the following commands:

```
library(ggplot2)
library(stars)
r <- read_stars(system.file("tif/L7_ETMs.tif", package = "stars"))
ggplot() + geom_stars(data = r) +
        facet_wrap(~band) + coord_equal() +
        theme_void() +
        scale_x_discrete(expand = c(0,0)) +
        scale_y_discrete(expand = c(0,0)) +
        scale_fill_viridis_c()
```

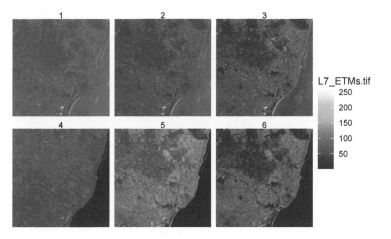

Figure 8.5: Simple facet raster plot with **ggplot2** and `geom_stars`

More elaborate **ggplot2**-based plots with **stars** objects may be obtained using package **ggspatial** (Dunnington 2022). Non-compatible but nevertheless **ggplot2**-style plots can be created with **tmap**, a package dedicated to creating high quality maps (Section 8.5).

When combining several feature sets with varying coordinate reference systems, using **geom_sf**, all sets are transformed to the reference system of the first set. To further control the "base" coordinate reference system, **coord_sf** can be used. This allows for instance working in a projected system, while combining graphics elements that are *not* **sf** objects but regular **data.frames** with ellipsoidal coordinates associated to WGS84.

8.5 Maps with tmap

Package **tmap** (Tennekes 2022, 2018) takes a fresh look at plotting spatial data in R. It resembles **ggplot2** in the sense that it composes graphics objects before printing by building on the **grid** package, and by concatenating map elements with a + between them, but otherwise it is entirely independent from, and incompatible with, **ggplot2**. It has a number of options that allow for highly professional looking maps, and many defaults have been carefully chosen. Creating a map with two similar attributes can be done using **tm_polygons** with two attributes, we can use

```
library(tmap)
system.file("gpkg/nc.gpkg", package = "sf") |>
    read_sf() |> st_transform('EPSG:32119') -> nc.32119
tm_shape(nc.32119) +
    tm_polygons(c("SID74", "SID79"), title = "SIDS") +
```

```
tm_layout(legend.outside = TRUE,
          panel.labels = c("1974-78", "1979-84")) +
tm_facets(free.scales=FALSE)
```

to create Figure 8.6:

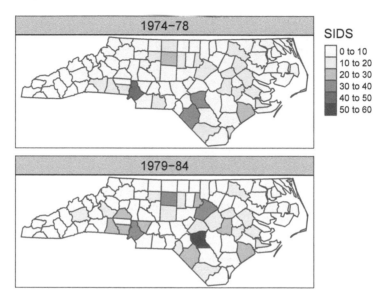

Figure 8.6: **tmap**: using `tm_polygons()` with two attribute names

Alternatively, from the long table form obtained by **pivot_longer** one could use +
`tm_polygons("SID") + tm_facets(by = "name")`.

Package **tmap** also has support for **stars** objects, an example created with

```
tm_shape(r) + tm_raster()
```

is shown in Figure 8.7. More examples of the use of **tmap** are given in Chapters
-Chapter 14 - -Chapter 16.

8.6   Interactive maps: `leaflet`, `mapview`, `tmap`

Interactive maps as shown in Figure 1.3 can be created with R packages **leaflet**,
mapview or **tmap**. **mapview** adds a number of capabilities to **leaflet** including a
map legend, configurable pop-up windows when clicking features, support for raster
data, and scalable maps with very large feature sets using the FlatGeobuf file format,

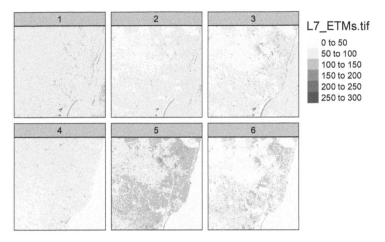

Figure 8.7: Simple raster plot with tmap

as well as facet maps that react synchronously to zoom and pan actions. Package **tmap** has the option that after giving

```
tmap_mode("view")
```

all usual **tmap** commands are applied to an interactive html/leaflet widget, whereas after

```
tmap_mode("plot")
```

again all output is sent to R's own (static) graphics device.

8.7 Exercises

1. For the countries Indonesia and Canada, create individual plots using equirectangular, orthographic, and Lambert equal area projections, while choosing projection parameters sensible for the area.
2. Recreate the plot in Figure 8.3 with **ggplot2** and with **tmap**.
3. Recreate the plot in Figure 8.7 using the **viridis** colour ramp.
4. View the interactive plot in Figure 8.7 using the "view" (interactive) mode of **tmap**, and explore which interactions are possible; also explore adding + tm_facets(as.layers=TRUE) and try switching layers on and off. Try also setting a transparency value to 0.5.

9

Large data and cloud native

This chapter describes how large spatial and spatiotemporal datasets can be handled with R, with a focus on packages **sf** and **stars**. For practical use, we classify large datasets as too large

- to fit in working memory,
- to fit on the local hard drive, or
- to download to locally managed infrastructure (such as network attached storage)

These three categories may (today) correspond very roughly to Gigabyte-, Terabyte- and Petabyte-sized datasets. Besides size considerations, access and processing speed also play a role, in particular for larger datasets or interactive applications. Cloud native geospatial formats are formats optimised with processing on cloud infrastructure in mind, where costs of computing and storage need to be considered and optimised. Such costs can be reduced by

- using compression, such as the LERC (Limited Error Raster Compression) algorithm used for cloud-optimised GeoTIFF, or the BLOSC compressor used for ZARR array data
- fast access to spatial sub-regions, such as by ways of HTTP range request enabled for cloud-optimised geotiffs, or column-oriented data access in the GeoParquet and GeoArrow formats
- accessing/viewing data at incremental resolution (low-resolution images are available before or without the full resolution being read), implemented natively by progressive JPEG formats or by using image pyramids (or overviews) in other raster formats
- optimising data access with particular cloud storage or object storage protocols in mind.

It should be noted that there are no silver bullets in this area: optimising storage for one particular access pattern will lead to slow access for other ways. If raster data is for instance stored for optimal access of spatial regions at different spatial resolutions, reading the data as pixel time series may be very slow. Compression leads to low storage and bandwidth costs but to higher processing costs when reading, because of the decompression involved.

9.1 Vector data: sf

9.1.1 Reading from local disk

Function **st_read** reads vector data from disk, using GDAL, and then keeps the data read in working memory. In case the file is too large to be read in working memory, several options exist to read parts of the file. The first is to set argument **wkt_filter** with a WKT text string containing a geometry; only geometries from the target file that intersect with this geometry will be returned. An example is

```
library(sf)
# Linking to GEOS 3.11.1, GDAL 3.6.2, PROJ 9.1.1; sf_use_s2()
# is TRUE
file <- system.file("gpkg/nc.gpkg", package = "sf")
c(xmin = -82,ymin = 36, xmax = -80, ymax = 37) |>
    st_bbox() |> st_as_sfc() |> st_as_text() -> bb
read_sf(file, wkt_filter = bb) |> nrow() # out of 100
# [1] 17
```

Here, **read_sf** is used as an alternative to **st_read** to suppress output.

The second option is to use the **query** argument to **st_read**, which can be any query in "OGR SQL" dialect. This can for instance be used to select features from a layer, or limit fields. An example is:

```
q <- "select BIR74,SID74,geom from 'nc.gpkg' where BIR74 > 1500"
read_sf(file, query = q) |> nrow()
# [1] 61
```

Note that **nc.gpkg** is the *layer name*, which can be obtained from **file** using **st_layers**. Sequences of records can be read using **LIMIT** and **OFFSET**, to read records 51-60 use

```
q <- "select BIR74,SID74,geom from 'nc.gpkg' LIMIT 10 OFFSET 50"
read_sf(file, query = q) |> nrow()
# [1] 10
```

Further query options include selection on geometry type or polygon area. When the dataset queried is a spatial database, then the query is passed on to the database and not interpreted by GDAL; this means that more powerful features will be available. Further information is found in the GDAL documentation under "OGR SQL dialect".

Very large files or directories that are zipped can be read without the need to unzip them, using the **/vsizip** (for zip), **/vsigzip** (for gzip), or **/vsitar** (for tar files) prefix to files. This is followed by the path to the zip file, and then followed by the file inside this zip file. Reading files this way may come at some computational cost.

9.1.2   Reading from databases, dbplyr

Although GDAL has support for several spatial databases, and as mentioned above it passes on SQL in the **query** argument to the database, it is sometimes beneficial to directly read from and write to a spatial database using the DBI R database drivers for this. An example of this is:

```
pg <- DBI::dbConnect(
    RPostgres::Postgres(),
    host = "localhost",
    dbname = "postgis")
read_sf(pg, query =
        "select BIR74,wkb_geometry from nc limit 3") |> nrow()
# [1] 3
```

A spatial query might look like

```
q <- "SELECT BIR74,wkb_geometry FROM nc WHERE \
ST_Intersects(wkb_geometry, 'SRID=4267;POINT (-81.50 36.43)');"
read_sf(pg, query = q) |> nrow()
# [1] 1
```

Here, the intersection is done in the database, and uses the spatial index when present. The same mechanism works when using **dplyr** with a database backend:

```
library(dplyr, warn.conflicts = FALSE)
nc_db <- tbl(pg, "nc")
```

Spatial queries can be formulated and are passed on to the database:

```
nc_db |>
    filter(ST_Intersects(wkb_geometry,
                    'SRID=4267;POINT (-81.50 36.43)')) |>
    collect()
# # A tibble: 1 x 16
#   ogc_fid  area perimeter cnty_ cnty_id name  fips  fipsno
#     <int> <dbl>     <dbl> <dbl>   <dbl> <chr> <chr>  <dbl>
# 1       1 0.114      1.44  1825    1825 Ashe  37009  37009
# # ... with 8 more variables: cress_id <int>, bir74 <dbl>,
# #   sid74 <dbl>, nwbir74 <dbl>, bir79 <dbl>, sid79 <dbl>,
# #   nwbir79 <dbl>, wkb_geometry <pq_gmtry>
nc_db |> filter(ST_Area(wkb_geometry) > 0.1) |> head(3)
# # Source:    SQL [3 x 16]
# # Database: postgres [edzer@localhost:5432/postgis]
#   ogc_fid  area perimeter cnty_ cnty_id name     fips  fipsno
#     <int> <dbl>     <dbl> <dbl>   <dbl> <chr>    <chr>  <dbl>
# 1       1 0.114      1.44  1825    1825 Ashe     37009  37009
```

```
# 2        3 0.143      1.63  1828     1828 Surry       37171  37171
# 3        5 0.153      2.21  1832     1832 Northamp~ 37131  37131
# # ... with 8 more variables: cress_id <int>, bir74 <dbl>,
# #   sid74 <dbl>, nwbir74 <dbl>, bir79 <dbl>, sid79 <dbl>,
# #   nwbir79 <dbl>, wkb_geometry <pq_gmtry>
```

(Note that PostGIS' **ST_Area** computes the same area as the **AREA** field in **nc**, which is the meaningless value obtained by assuming the coordinates are projected, where they are ellipsoidal.)

9.1.3   Reading from online resources or web services

GDAL drivers support reading from online resources, by prepending the URL starting with **https://** with **/vsicurl/**. A number of similar drivers specialised for particular clouds include **/vsis3/** for Amazon S3, **/vsigs/** for Google Cloud Storage, **/vsiaz/** for Azure, **/vsioss/** for Alibaba Cloud, or **/vsiswift/** for OpenStack Swift Object Storage. Section 9.3.2 has an example of using **/vsicurl/**.

These prepositions can be combined with **/vsizip/** to read a file from a zipped online resource. Depending on the file format used, reading information this way may require reading the entire file, or reading it multiple times, and may not always be the most efficient way of handling resources. Cloud native formats are optimised to work efficiently using **HTTP** requests.

9.1.4   APIs, OpenStreetMap

Typical web services for geospatial data create data on the fly and give access to this through an API. As an example, data from OpenStreetMap can be bulk downloaded and read locally, for instance using the GDAL vector driver, but more typical a user wants to obtain a small subset of the data or use the data for a small query. Several R packages exist that query OpenStreetMap data:

- Package **OpenStreetMap** (Fellows and JMapViewer library by Jan Peter Stotz 2019) downloads data as raster tiles, typically used as backdrop or reference for plotting other features
- Package **osmdata** (Mark Padgham et al. 2017) downloads vector data as points, lines, or polygons in **sf** or **sp** format
- Package **osmar** (from CRAN archive) returns vector data, but in addition the network topology (as an **igraph** object) that contains how road elements are connected, and has functions for computing the shortest route

When provided with a correctly formulated API call in the URL the highly configurable GDAL OSM driver (in **st_read**) can read an ".osm" file (xml) and return a dataset with five layers: **points** that have significant tags, **lines** with non-area "way" features, **multilinestrings** with "relation" features, **multipolygons** with "relation" features , and **other_relations**. A simple and very small bounding box query to OpenStreetMap could look like

```
download.file(paste0("https://openstreetmap.org/api/0.6/map?",
        "bbox=7.595,51.969,7.598,51.970"),
    "data/ms.osm", method = "auto")
```

and from this file we can read the layer **lines**, and plot its first attribute by

```
o <- read_sf("data/ms.osm", "lines")
p <- read_sf("data/ms.osm", "multipolygons")
st_bbox(c(xmin = 7.595, ymin = 51.969,
          xmax = 7.598, ymax = 51.970), crs = 'OGC:CRS84') |>
    st_as_sfc() |>
    plot(axes = TRUE, lwd = 2, lty = 2, cex.axis = .5)
plot(o[,1], lwd = 2, add = TRUE)
plot(st_geometry(p), border = NA, col = '#88888888', add = TRUE)
```

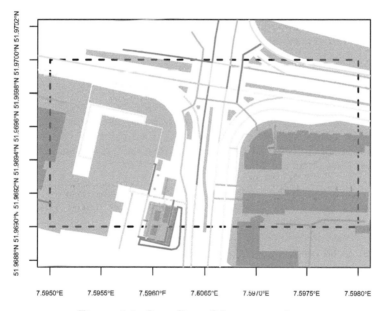

Figure 9.1: OpenStreetMap vector data

the result of which is shown in Figure 9.1. The overpass API provides a more generic
and powerful query functionality to OpenStreetMap data.

9.1.5 GeoParquet and GeoArrow

Two formats dedicated to cloud native analysis are derived from the Apache projects
Parquet and Arrow. Both provide column oriented storage of tabular data, meaning
that the reading of a single field for many records is fast, compared to record oriented
storage of most other databases. The Geo- extensions of both involve

- a way of storing a geometry column, either in a well-known binary or text form, or in a more efficient form where sub-geometries are indexed up front
- storage of a coordinate reference system.

At the time of writing this book, both formats are under active development, but drivers for reading or creating them are available in GDAL starting from version 3.5. Both formats allow for compressed storage. The difference is that (Geo)Parquet is more oriented towards persistent storage, where, (Geo)Arrow is more oriented to fast access and fast computation. Arrow can for instance be both an in-memory and an on-disk format.

9.2   Raster data: stars

A common challenge with raster datasets is not only that they come in large files (single Sentinel-2 tiles are around 1 GB), but that many of these files, potentially thousands or millions, are needed to address the area and time period of interest. In 2022, Copernicus, the program that runs all Sentinel satellites, published 160 TB of images per day. This means that a classic pattern in using R consisting of downloading data to local disc, loading the data in memory, and analysing it is not going to work.

Cloud-based Earth Observation processing platforms like Google Earth Engine (Gorelick et al. 2017), Sentinel Hub or openEO.cloud recognise this and let users work with datasets up to the petabyte range rather easily and with a great deal of interactivity. They share the following properties:

- computations are postponed as long as possible (lazy evaluation)
- only the data asked for by the user is being computed and returned, and nothing more
- storing intermediate results is avoided in favour of on-the-fly computations
- maps with useful results are generated and shown quickly to allow for interactive model development

This is similar to the **dbplyr** interface to databases and cloud-based analytics environments, but differs in the aspect of *what* we want to see quickly: rather than the first n records of a **dbplyr** lazy table, we want a quick *overview* of the results, in the form of a map covering the whole area, or part of it, but at screen resolution rather than native (observation) resolution.

If for instance we want to "see" results for the United States on a screen with 1000 × 1000 pixels, we only need to compute results for this many pixels, which corresponds roughly to data on a grid with 3000 m × 3000 m grid cells. For Sentinel-2 data with 10 m resolution, this means we can downsample with a factor 300, giving 3 km × 3 km resolution. Processing, storage, and network requirements then drop a factor $300^2 \approx 10^5$, compared to working on the native 10 m × 10 m resolution. On the platforms mentioned, zooming in the map triggers further computations on a finer resolution and smaller extent.

A simple optimisation that follows these lines is how the plot method for **stars** objects works. In the case of plotting large rasters, it downsamples the array before it plots, drastically saving time. The degree of downsampling is derived from the plotting region size and the plotting resolution (pixel density). For vector devices, such as pdf, R sets plot resolution to 75 dpi, corresponding to 0.3 mm per pixel. Enlarging plots may reveal this, but replotting to an enlarged devices will create a plot at target density. For **geom_stars** the user has to specify the **downsample** rate, because the **ggplot2** does not make the device size available to that function.

9.2.1   stars proxy objects

To handle datasets that are too large to fit in memory, **stars** provides **stars_proxy** objects. To demonstrate its use, we will use the **starsdata** package, an R data package with larger datasets (around 1 GB total). It can be installed by

```
options(timeout = 600) # or large in case of slow network
install.packages("starsdata",
    repos = "http://pebesma.staff.ifgi.de", type = "source")
```

We can "load" a Sentinel-2 image from it by

```
library(stars) |> suppressPackageStartupMessages()
f <- paste0("sentinel/S2A_MSIL1C_20180220T105051_N0206",
            "_R051_T32ULE_20180221T134037.zip")
granule <- system.file(file = f, package = "starsdata")
file.size(granule)
# [1] 7.69e+08
base_name <- strsplit(basename(granule), ".zip")[[1]]
s2 <- paste0("SENTINEL2_L1C:/vsizip/", granule, "/", base_name,
    ".SAFE/MTD_MSIL1C.xml:10m:EPSG_32632")
(p <- read_stars(s2, proxy = TRUE))
# stars_proxy object with 1 attribute in 1 file(s):
# $EPSG_32632
# [1] "[...]/MTD_MSIL1C.xml:10m:EPSG_32632"
#
# dimension(s):
#         from     to offset delta           refsys    values x/y
# x          1 10980   3e+05    10 WGS 84 / UTM z...     NULL [x]
# y          1 10980   6e+06   -10 WGS 84 / UTM z...     NULL [y]
# band       1     4      NA    NA                  NA B4,...,B8
object.size(p)
# 12576 bytes
```

and we see that this does not actually load *any* of the pixel values but keeps the reference to the dataset and fills the dimensions table. (The convoluted **s2** name is needed to point GDAL to the right file inside the **.zip** file containing 115 files in total).

The idea of a proxy object is that we can build expressions like

```
p2 <- p * 2
```

but that the computations for this are postponed. Only when we really need the data, is `p * 2` evaluated. We need data when we:

- `plot` data,
- write an object to disk, with `write_stars`, or
- explicitly load an object in memory, with `st_as_stars`

In case the entire object does not fit in memory, `plot` and `write_stars` choose different strategies to deal with this:

- `plot` fetches only the pixels that can be seen by downsampling, rather than reading all pixels available
- `write_stars` reads, processes, and writes data chunk-by-chunk

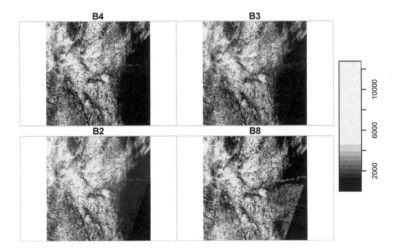

Figure 9.2: Downsampled 10 m bands of a Sentinel-2 scene

Downsampling and chunking is implemented for spatially dense images, but not for dense time series or other dense dimensions. As an example, the output of `plot(p)`, shown in Figure 9.2 only fetches the pixels that can be seen on the plot device, rather than the 10980 × 10980 pixels available in each band. The downsampling ratio taken is

```
# [1] 69
```

meaning that for every 69 × 69 sub-image in the original image, only one pixel is read and plotted.

9.2.2   Operations on proxy objects

Several dedicated methods are available for **stars_proxy** objects:

```
methods(class = "stars_proxy")
#  [1] [                 [[<-             [<-
#  [4] adrop             aggregate        aperm
#  [7] as.data.frame     c                coerce
# [10] dim               droplevels       filter
# [13] hist              initialize       is.na
# [16] Math              merge            mutate
# [19] Ops               plot             predict
# [22] print             pull             rename
# [25] select            show             slice
# [28] slotsFromS3       split            st_apply
# [31] st_as_sf          st_as_stars      st_crop
# [34] st_dimensions<-   st_downsample    st_mosaic
# [37] st_redimension    st_sample        st_set_bbox
# [40] transmute         write_stars
# see '?methods' for accessing help and source code
```

We have seen **plot** and **print** in action; **dim** reads out the dimension from the dimensions metadata table.

The three methods that actually fetch data are **st_as_stars**, **plot** and **write_stars**. **st_as_stars** reads the actual data into a **stars** object, its argument **downsample** controls the downsampling rate. **plot** does this too, choosing an appropriate **downsample** value from the device resolution, and plots the object. **write_stars** writes a **stars_proxy** object to disk.

All other methods for **stars_proxy** objects do not actually operate on the raster data but add the operations to a *to do* list attached to the object. Only when actual raster data are fetched, for instance when **plot** or **st_as_stars** are called, the commands in this list are executed.

st_crop limits the extent (area) of the raster that will be read. **c** combines **stars_proxy** objects, but still doesn't read any data. **adrop** drops empty dimensions and **aperm** can change dimension order.

write_stars reads and processes its input chunk-wise; it has an argument **chunk_size** that lets users control the size of spatial chunks.

9.2.3   Remote raster resources

A format like "cloud-optimised GeoTIFF" (COG) has been specially designed to be efficient and resource-friendly in many cases, e.g., for only reading the metadata, or for only reading overviews (low-resolution versions of the full imagery) or spatial regions using the **/vsixxx/** mechanisms (Section 9.1.3). COGs can also be created using the GeoTIFF driver of GDAL, and setting the right dataset creation options in a **write_stars** call.

9.3   Very large data cubes

At some stage, datasets need to be analysed that are so large that downloading them is no longer feasible; even when local storage would be sufficient, network bandwidth may become limiting. Examples are satellite image archives such as those from Landsat and Copernicus (Sentinel-x), or model computations such as the ERA5 (Hersbach et al. 2020), a model reanalysis of the global atmosphere, land surface, and ocean waves from 1950 onwards. In such cases it may be most helpful to gain access to virtual machines in a cloud that have these data available, or to use a system that lets the user carry out computations without having to worry about virtual machines and storage. Both options will be discussed.

9.3.1   Finding and processing assets

When working on a virtual machine on a cloud, a first task is usually to find the assets (files) to work on. It looks attractive to obtain a file listing, and then parse file names such as

```
S2A_MSIL1C_20180220T105051_N0206_R051_T32ULE_20180221T134037.zip
```

for their metadata including the date of acquisition and the code of the spatial tile covered. Obtaining such a file listing however is usually computationally very demanding, as is the processing of the result, when the number of tiles runs in the many millions.

A solution to this is to use a catalogue. The recently developed and increasingly deployed STAC, short for *spatiotemporal asset catalogue*, provides an API that can be used to query image collections by properties like bounding box, date, band, and cloud coverage. The R package **rstac** (Simoes, Carvalho, and Brazil Data Cube Team 2023) provides an R interface to create queries and manage the information returned.

Processing the resulting files may involve creating a data cube at a lower spatial and/or temporal resolution, from images that may span a range of coordinate reference systems (e.g., several UTM zones). An R package that creates a regular data cube from such a collection of images is **gdalcubes** (Appel 2023; Appel and Pebesma 2019), which can also directly use a STAC (Appel, Pebesma, and Mohr 2021) to identify images.

9.3.2   Cloud native storage: Zarr

Where COG provides cloud native storage for raster imagery, Zarr is a format for cloud native storage of large multi-dimensional arrays. It can be seen as a successor of NetCDF and seems to follow similar conventions, being used by the climate and forecast communities (Eaton et al. 2022). Zarr "files" are really directories with sub-directories containing compressed chunks of the data. The compression algorithm and the chunking strategy will have an effect on how fast particular sub-cubes can be read or written.

Function **stars::read_mdim** can read entire data cubes but has options for reading sub-cubes by specifying for each dimension offset, number of pixels, and the step size to read a dimension at a lower resolution (Pebesma 2022a). Similarly, **stars::write_mdim** can write multi-dimensional arrays to Zarr or NetCDF files, or other formats that support the GDAL C++ multi-dimensional array API.

To read a remote (cloud-based) Zarr file, one needs to prepend the URL with indicators about the format and the access protocol:

```
dsn = paste0('ZARR:"/vsicurl/https://ncsa.osn.xsede.org',
        '/Pangeo/pangeo-forge/gpcp-feedstock/gpcp.zarr"')
```

after which we can read the first 10 time steps using

```
library(stars)
bounds = c(longitude = "lon_bounds", latitude = "lat_bounds")
(r = read_mdim(dsn, bounds = bounds, count = c(NA, NA, 10)))
# stars object with 3 dimensions and 1 attribute
# attribute(s):
#               Min. 1st Qu. Median Mean 3rd Qu. Max.
# precip [mm/d]    0       0      0 2.25     1.6  109
# dimension(s):
#             from  to          offset delta refsys x/y
# longitude      1 360               0     1     NA [x]
# latitude       1 180             -90     1     NA [y]
# time           1  10 1996-10-01 UTC 1 days POSIXct
st_bbox(r)
# xmin ymin xmax ymax
#    0  -90  360   90
```

In this example,

- **NA** values for **count** indicate to get all values available for that dimension
- here, **bounds** variables needed explicit specification because the data source did not follow the more recent CF (1.10) conventions; and ignoring bounds would lead to a raster with a bounding box having latitude values outside $[-90, 90]$.

9.3.3   APIs for data: GEE, openEO

Platforms that do not require the management and programming of virtual machines *in* the cloud but provide direct access to the imagery managed include GEE, openEO, and the climate data store.

Google Earth Engine (GEE) is a cloud platform that allows users to compute on large amounts of Earth Observation data as well as modelling products (Gorelick et al. 2017). It has powerful analysis capabilities, including most of the data cube operations explained in Section 6.3. It has an interface where scripts can be written in JavaScript, and a Python interface to the same functionality. The code of GEE is not open-source, and cannot be extended by arbitrary user-defined functions in languages like Python or R. R package **rgee** (Aybar 2022) provides an R client interface to GEE.

Cloud-based data cube processing platforms built entirely around open source software are emerging, several of which using the openEO API (Schramm et al. 2021). This API allows for user-defined functions (UDFs) written in Python or R that are being passed on through the API and executed at the pixel level, for instance to aggregate or reduce dimensions using custom reducers. UDFs in R represent the data chunk to be processed as a `stars` object; in Python `xarray` objects are used.

Other platforms include the Copernicus climate data store (Raoult et al. 2017) or atmosphere data store, which allow processing of atmospheric or climate data from ECMWF, including ERA5. An R package with an interface to both data stores is **ecmwfr** (Hufkens 2023).

9.4 Exercises

Use R to solve the following exercises.

1. For the S2 image (above), find out in which order the bands are by using `st_get_dimension_values()`, and try to find out which spectral bands / colours they correspond to.
2. Compute NDVI for the S2 image, using `st_apply` and an appropriate **ndvi** function. Plot the result to screen, and then write the result to a GeoTIFF. Explain the difference in runtime between plotting and writing.
3. Plot an RGB composite of the S2 image, using the **rgb** argument to `plot()`, and then by using `st_rgb()` first.
4. Select five random points from the bounding box of S2, and extract the band values at these points; convert the object returned to an **sf** object.
5. For the 10-km radius circle around `POINT(390000 5940000)`, use **aggregate** to compute the mean pixel values of the S2 image when downsampling the images with factor 30, and on the original resolution. Compute the relative difference between the results.
6. Use **hist** to compute the histogram on the downsampled S2 image. Also do this for each of the bands. Use **ggplot2** to compute a single plot with all four histograms.
7. Use `st_crop` to crop the S2 image to the area covered by the 10-km circle. Plot the results. Explore the effect of setting argument `crop = FALSE`
8. With the downsampled image, compute the logical layer where all four bands have pixel values higher than 1000. Use a raster algebra expression on the four bands (use `split` first), or use `st_apply` for this.

Part III

Models for Spatial Data

The first two parts of this book contain a considerable amount of concepts that one could classify as "models for spatial data", including:

- how numbers relate to real-world phenomena (Chapter 2)
- how coordinates are defined in different spaces (Chapter 2, Chapter 4)
- simple feature geometries (Chapter 3), how straight lines between points can be used to define linestrings and polygons
- the set of geometry types (Section 3.1)
- support and the way feature attributes can relate to geometries (Chapter 5)
- how simple tesselations can describe space-time volumes (Chapter 6)
- how these concepts can be made operational using data science software (Chapter 7)

The third and largest part of this book is dedicated to *statistical* modelling of spatial data. The scientific discipline *statistics* is concerned with describing and understanding variability in observations, and predicting future observations. Observations are often modelled as

$$\text{observed} = \text{explained} + \text{remainder}$$

where "remainder" refers to variation that could not be explained by predictors, including measurement error but also lack of fit or variation caused by model misspecification. For spatial data, a further term is often helpful, as in

$$\text{observed} = \text{explained} + \text{smooth} + \text{remainder}$$

where "smooth" refers to variation that is not explained by external predictors but that clearly shows "smooth" spatial patterns, as opposed to the "rough" remainder which does not do this. Such a "smooth" term can for instance be modelled by base functions in coordinates (splines, smoothers) or as a random term that is spatially correlated.

Chapter 10 introduces statistical modelling of spatial data, as a preparation to the subsequent chapters but also highlighting a number of relevant aspects that are not elaborated on in later chapters. It tries to bridge these chapters with concepts from the first part of this book, in particular support (Chapter 5).

It is now obvious that a complete and comprehensive treatment of the topic of statistical models for spatial data that also includes instructions about the use of computational software in a single book is an impossible task. The `spatstat` book (Baddeley, Rubak, and Turner 2015) has around 850 pages for only spatial point patterns and R. This part focuses on the three "classical" spatial statistics topics: analysis of point patterns (Chapter 11), geostatistical data (Chapters -Chapter 12 and -Chapter 13), and lattice (areal) data (Chapters -Chapter 14 - -Chapter 17). Where possible we attempt to refer to further literature on methods and software implementations in R.

10

Statistical modelling of spatial data

So far in this book, we mostly addressed the problem of *describing* data. This included geometrical measures, predicates, or transformations that involved geometries, or by summary measures of attributes, or by plots involving variability in the geometry, the feature attributes, or both.

Statistical modelling aims at going beyond describing the data, it considers the data as a sample drawn from a population, and tries to make assessments (inference) about the population sampled from, for instance by quantifying relationships between variables, estimating population parameters, and predicting the outcome of observations that could have been taken but were not, as is the case in interpolation problems. This is usually done by adopting a model for the data, where for instance observations are decomposed as follows:

$$\text{observed} = \text{explained} + \text{remainder} \tag{10.1}$$

where "explained" typically uses external variables (predictors, covariates, in machine learning confusingly also called features) that are related to the observed variable and some kind of regression model to translate into variability of the observed variable, and "remainder" is remaining variability that could not be explained. Interest may focus on the nature and magnitude or the relations between predictors and the observed variable, or in predicting new observations.

Statistical models, and *sampling* hinge on the concept of probability, which in typical spatial data science problems is not a force of nature but has to be assumed in one way or another. If we are faced with data that come from (spatially) random sampling and we are interested in estimating means or totals, a *design-based* approach that assumes randomness in the sample locations is the most straightforward analysis approach, as pointed out in more detail in Section 10.4. If observations were not sampled randomly, or if our interest is in predicting values at specific locations (mapping), a *model-based* approach is needed. The remaining chapters in this part deal with model-based approaches.

10.1 Mapping with non-spatial regression and ML models

Regression models or other machine learning (ML) models can be applied to spatial and spatiotemporal data just the way they are applied for predicting new observations in non-spatial problems:

1. **estimate**: for a set of observations, a regression or ML model is fitted using predictor values corresponding to the observations (in ML jargon, this step is also known as "train")
2. **predict**: for a new situation, known predictor values are combined with the fitted model to predict the value of the observed variable, along with a prediction error or prediction interval if possible

Objects of class **sf** need no special treatment, as they are **data.frames**. To create maps of the resulting predictions, predicted values need to be added to the **sf** object, which can be done using the **nc** dataset loaded as in Chapter 1 by:

```
nc |> mutate(SID = SID74/BIR74, NWB = NWBIR74/BIR74) -> nc1
lm(SID ~ NWB, nc1) |>
  predict(nc1, interval = "prediction") -> pr
bind_cols(nc, pr) |> names()
#  [1] "AREA"      "PERIMETER" "CNTY_"     "CNTY_ID"
#  [5] "NAME"      "FIPS"      "FIPSNO"    "CRESS_ID"
#  [9] "BIR74"     "SID74"     "NWBIR74"   "BIR79"
# [13] "SID79"     "NWBIR79"   "geom"      "fit"
# [17] "lwr"       "upr"
```

where we see that

- **lm** estimates a linear model and works directly on an **sf** object
- the output is used for a **predict** model, which predicts values corresponding to the observations in **nc1**, the same **sf** object
- **predict** creates three columns: **fit** for predicted values and **lwr** and **upr** for the 95% prediction intervals
- these three columns have been added to the final object using **bind_cols**.

In general the datasets for model estimation and prediction do not have to be the same. Section 7.4.6 points out how this can be done with **stars** objects (essentially by going through a long **data.frame** representation of the datacube and converting the predicted results back, potentially in a chunked fashion).

Because many regression and ML type problems share this same structure, packages like **caret** (Kuhn 2022) or **tidymodels** (Kuhn and Wickham 2022) allow for automated evaluation and comparison over a large set of model alternatives, offering a large set of model evaluation criteria and cross-validation strategies. Such cross-validation approaches assume independent observations, which is often not a reasonable assumption for spatial data, for instance because of spatial correlation (Ploton et al. 2020) or because of strong spatial clustering in sample data (Meyer and Pebesma 2022), or both, and a number of R packages provide methods that are meant as replacements for naive cross-validation, including **spatialsample** (Silge and Mahoney 2023), **CAST** (Meyer, Milà, and Ludwig 2023), **mlr3spatial** (Becker and Schratz 2022), and **mlr3spatiotempcv** (Schratz and Becker 2022).

Strong spatial clustering of sample can arise when sample data are composed by joining different databases, each with very different sampling density. This is often the case in global datasets (Meyer and Pebesma 2022). Another example of strong clustering arises when, for sampling ground truth points of a land cover class, polygons are digitised and points are sampled within these polygons at the resolution of pixels in satellite imagery.

Spatial correlation in the "remainder" part of the model may be decreased by adding spatial coordinates or functions of spatial coordinates to the set of predictors. This also carries a risk of over-optimistic predictions in extrapolation cases, (cross-) validation, and model assessment, and is further discussed in Section 10.5.

10.2   Support and statistical modelling

Support of data (Section 1.6; Chapter 5) plays a lead role in the statistical analysis of spatial data. Methods for areal data (Chapters -Chapter 14 - -Chapter 17) are devised for data with area support, where the set of areas cover the entire area of interest.

By showing an extensive variable (Section 5.3.1) in a polygon choropleth map as done in Figure 1.1 one runs the risk that the information is related with the polygon size, and that the signal shown is actually the size of the polygons, in colour. For the variable *population count* one would divide by the polygon area to show the (intensive) variable *population density*, in order to create an informative map. In the analysis of health data, like disease incidences over a time period shown in Figure 1.1, rather than dividing by polygon area to get a spatial density, observations are usually converted to probabilities or *incidence rates* by dividing over the population size of the associated polygons. As such they are (still) associated with the polygon area but their support is associated with the population total. It is these totals that inform the (Poisson) variability used by subsequent modelling in for instance CAR-type models (Chapter 16).

Chapter 11 deals in principle with point support observations, but at some stage needs to acknowledge that observations have non-zero size: tree stem "points" cannot be separated distances smaller than the tree diameter. Also, points in point pattern analysis are considered in their *observation window*, the area for which the point dataset is exhaustive, or complete. The observation window is of influence in many of the analysis tools. If points are observed on a line network, then the observation window consists of the set of lines observed, and distances measured through this network.

Geostatistical data (Chapters -Chapter 12 and -Chapter 13) usually start with point support observations and may end with predictions (spatial interpolations) for unobserved point locations distributed over the area of interest, or, may end in predictions for means over areas (block kriging; Section 12.5). Alternatively, observations may be aggregates over regions (Skøien et al. 2014). In remote sensing data, pixel values are usually associated with aggregates over the pixel area. Challenges may be the filling of gaps in images such as gaps caused by cloud coverage, from pixels neighbouring in space and time Militino et al. (2019).

When combining data with different spatial supports, for instance polygons from administrative regions and raster layers, it is often seen that all information is "brought together" to the highest resolution, by simply extracting polygon values at pixel locations, and proceeding from there, with all the newly created "observations". This of course bears a large risk of producing non-sensible results when analysing these "data", and a proper downsampling strategy, possibly using simulations to cope with uncertainty, would be a better alternative. For naive users, using software that is not aware of support of values associated with areas and using software that does not warn against naive downsampling is of course not a helpful situation.

10.3   Time in predictive models

Schabenberger and Gotway (2005) already noted that in many cases, statistical analysis of spatiotemporal data proceeds either by reducing time, then working on the problem spatially (time first, space later) or reducing space, then working on the problem temporally (space first, time later). An example of the first approach is given in Chapter 12 where a dataset with a year of hourly values (detailed in Chapter 13) are reduced to station mean values (time first) after which these means are interpolated spatially (space later). Examples from the area of remote sensing are

- Simoes et al. (2021) use supervised machine learning and time series deep learning to segmentise pixel time series into sequences of land use (time first), and then smooth the resulting sequences of maps to remove improbable transitions in isolated pixels (space later)
- Verbesselt et al. (2010) use (unsupervised) structural change algorithms to find breakpoints in pixel time series (time first), which are interpreted in the context deforestation later on.

Examples of space first, time later in the area of remote sensing are any case where a single scene or scenes belonging to a single season are classified, and multi-year changes in land use or land cover are assessed by comparing time sequences of classified scenes. An example of this is C. F. Brown et al. (2022). Examples where space and time are considered *jointly* are the spatiotemporal interpolation in Chapter 13, and Lu et al. (2016) in the context of remote sensing.

10.4 Design-based and model-based inference

Statistical inference means the action of estimating parameters about a population from sample data. Suppose we denote the variable of interest with $z(s)$, where z is the attribute value measured at location s, and we are interested in estimating the mean value of $z(s)$ over a domain D,

$$z(s) = \frac{1}{|D|} \int_{u \in D} z(u)du,$$

with $|D|$ the area of D, from sample data $z(s_1), ..., z(s_n)$.

Then, there are two possibilities to proceed: model-based, or design-based. A model-based approach considers $z(s)$ to be a realisation of a super-population $Z(s)$ (using capital letters to indicate random variables), and could for instance postulate a model for its spatial variability in the form of

$$Z(s) = m + e(s), \quad \mathrm{E}(e(s)) = 0, \quad \mathrm{Cov}(e(s)) = \Sigma(\theta)$$

with m a constant mean and $e(s)$ a residual with mean zero and covariance matrix $\Sigma(\theta)$. This would require choosing the covariance function $\Sigma()$ and estimating its parameters θ from $z(s)$, and then computing a block kriging prediction $\hat{Z}(D)$ (Section 12.5). This approach makes no assumptions about how $z(s)$ was sampled *spatially*, but of course it should allow for choosing the covariance function and estimating its parameters; inference is conditional to the validity of the postulated model.

Rather than assuming a superpopulation model, the design-based approach (De Gruijter and Ter Braak 1990; Brus 2021a; Breidt, Opsomer, et al. 2017) assumes randomness in the locations, which is justified (only) when using random sampling. It *requires* that the sample data were obtained by probability sampling, meaning that some form of spatial random sampling was used where all elements of $z(s)$ had a known and positive probability of being included in the sample obtained. The random process is that of sampling: $z(s_1)$ is a realisation of the random process $z(S_1)$, the first observation taken *over repeated random sampling*. Design-based estimators only need these inclusion probabilities to estimate mean values with standard errors. This means that for instance given a simple random sample, the unweighted sample mean is used to estimate the population mean, and no model parameters need to be fit.

Now the question is whether $z(s_1)$ and $z(s_2)$ can be expected to be correlated when s_1 and s_2 are close together. The question does not work out as long as $z(s_1)$ and $z(s_2)$ are just two numbers: we need some kind of framework, random variables, that recreates this situation to form two sets of numbers for which we can consider correlation. The misconception here, as explained in Brus (2021a), is that the two are always spatially correlated, but this is only the case when working under model-based approaches: $Z(s_1)$ and $Z(s_2)$ may well be correlated ("model-dependent"), but although in a particular random sample (realisation) $z(s_1)$ and $z(s_2)$ *may* be close in space, the corresponding random variables $z(S_1)$ and $z(S_2)$ considered over repeated random sampling are not close together, and, are design-independent. Both situations can coexist without contradiction and are a consequence of choosing to work under one inference framework or the other.

The choice whether to work under a design-based or model-based framework depends on the purpose of the study and the data collection process. The model-based framework lends itself best for cases:

- where predictions are required for individual locations, or for areas too small to be sampled
- where the available data were not collected using a known random sampling scheme (i.e., the inclusion probabilities are unknown, or are zero over particular areas and or times)

Design-based approaches are most suitable when:

- observations were collected using a spatial random sampling process
- aggregated properties of the entire sample region (or sub-region) are needed
- estimates are required that are not sensitive to model misspecification, for instance when needed for regulatory or legal purposes

In case a sampling procedure is to be planned (De Gruijter et al. 2006), some form of spatial random sampling is definitely worth considering since it opens up the possibility of following both inference frameworks.

10.5   Predictive models with coordinates

In data science projects, coordinates may be seen as features in a larger set of predictors (or features, or covariates) and treated accordingly. There are some pitfalls in doing so.

As usual when working with predictors, it is good to choose predictive methods that are not sensitive to shifts in origin or shifts in unit (scale). Assuming a two-dimensional problem, predictive models should also not be sensitive to arbitrary rotations of the x- and y- or latitude and longitude axes. For projected (2D, Cartesian) coordinates this can be assured, e.g., by using polynomials of order n as $(x + y)^n$, rather than $(x)^n + (y)^n$; for a second order polynomial this involves including the term xy, so that an ellipsoidal-shape trend surface does not have to be aligned with the $x-$ or $y-$axis. For a GAM model with spline components, one would use a spline in two dimensions $s(x, y)$ rather than two independent splines $s(x)$ and $s(y)$ that do not allow for interaction. An exception to this "rule" would be when a pure latitude effect is desired for instance to account for yearly total solar energy influx.

When the area covered by the data is large, the difference between using ellipsoidal coordinates and projected coordinates will automatically become larger, and hence choosing one of both will have an effect on predictive modelling. For very large extents and global models, polynomials or splines in latitude and longitude will not work well as they ignore the circular nature of longitude and the coordinate singularities at the poles. Here, spherical harmonics, base functions that are continuous on the sphere with increasing spatial frequencies can replace polynomials or be used as spline base functions.

In many cases, the spatial coordinates over which samples were collected also define the space over which predictions are made, setting them apart from other features. Many simple predictive approaches, including most machine learning methods, assume sample data to be independent. When samples are collected by spatially random sampling over the spatial target area, this assumption may be justified when working under a design-based context (Brus 2021b). This context, however, treats the coordinate space as the variable over which we randomise, which affords predicting values for a new *randomly chosen* location but rules out making predictions for fixed locations; this implies that averages of areas over which samples were collected, can be obtained, but not spatial interpolations. In case predictions for fixed locations are required, or in case data were not collected by spatial random sampling, a model-based approach (as taken in Chapter 12) is needed and typically some form of spatial and/or temporal autocorrelation of residuals must be assumed.

A common case is where sample data are collected opportunistically ("whatever could be found"), and are then used in a predictive framework that does not weight them. This has a consequence that the resulting model may be biased towards over-represented areas (in predictor space and/or in spatial coordinates space), and that simple (random) cross-validation statistics may be over-optimistic when taken as performance measures for spatial prediction (Meyer and Pebesma 2021, 2022; Mila et al. 2022). Adaptive cross-validation measures such as spatial cross-validation may help getting more relevant measures for predictive performance.

10.6   Exercises

Use R to solve the following exercises.

1. Following the `lm` example of Section 10.1 use a random forest model to predict `SID` values (e.g., using package **randomForest**), and plot the random forest predictions against observations, along with the $x = y$ line.
2. Create a new dataset by randomly sampling 1000 points from the **nc** dataset, and rerun the linear regression model of Section 10.1 on this dataset. Consider the `summary` of the fitted models, in particular the estimated coefficients, their standard errors, and the residual standard error. What has changed?

3. Redo the water-land classification of Section 7.4.6 using `class::knn` instead of `lda`, using a value of `k = 5`, and compare the resulting predictions with those of `lda`.

4. For the linear model using `nc` and for the `knn` example of the previous exercise, add a first and a second order linear model in the spatial coordinates and compare the results (use `st_centroid` to obtain polygon centroids, and `st_coordinates` to extract the `x` and `y` coordinates in matrix form).

11

Point Pattern Analysis

Point pattern analysis is concerned with describing patterns of points over space and making inference about the process that could have generated an observed pattern. The main focus here lies on the information carried in the locations of the points, and typically these locations are not controlled by sampling but a result of a process we are interested in, such as animal sightings, accidents, disease cases, or tree locations. This is opposed to geostatistical processes (Chapter 12) where we have values of some phenomenon everywhere but observations limited to a set of locations *that we can control*, at least in principle. Hence, in geostatistical problems the prime interest is not in the observation locations but in estimating the value of the observed phenomenon at unobserved locations. Point pattern analysis typically assumes that for an observed area, all points are available, meaning that locations without a point are not unobserved as in a geostatistical process, but are observed and contain no point. In terms of random processes, in point processes locations are random variables, where, in geostatistical processes the measured variable is a random field with locations fixed.

This chapter is confined to describing the very basics of point pattern analysis, using package **spatstat** (Baddeley, Turner, and Rubak 2022), and related packages by the same authors. The **spatstat** book of Baddeley, Rubak, and Turner (2015) gives a comprehensive introduction to point pattern theory and the use of the **spatstat** package family, which we will not try to copy. Inclusion of particular topics in this chapter should not be seen as an expression that these are more relevant than others. In particular, this chapter tries to illustrate interfaces existing between **spatstat** and the more spatial data science oriented packages **sf** and **stars**. A further book that introduces point patterns analysis is Stoyan et al. (2017). R package **stpp** (Gabriel et al. 2022) for analysing spatiotemporal point processes is discussed in Gabriel, Rowlingson, and Diggle (2013).

Important concepts of point patterns analysis are the distinction between a point *pattern* and a point *process*: the latter is the stochastic process that, when sampled, generates a point pattern. A dataset is always a point pattern, and inference involves figuring out the properties of a process that could have generated a pattern like the one we observed. Properties of a spatial point process include

- first order properties: the intensity function measures the number of points per area unit; this function is spatially varying for a *inhomogeneous* point process
- second order properties: given a constant or varying intensity function, describe whether points are distributed independently *from one another*, tend to attract each other (clustering), or repulse each other (more regularly distributed than under complete spatial randomness)

11.1   Observation window

Point patterns have an observation window. Consider the points generated randomly
by

```
library(sf)
# Linking to GEOS 3.11.1, GDAL 3.6.2, PROJ 9.1.1; sf_use_s2()
# is TRUE
n <- 30
set.seed(13531) # remove this to create another random sequence
xy <- data.frame(x = runif(n), y = runif(n)) |>
    st_as_sf(coords = c("x", "y"))
```

then these points are (by construction) uniformly distributed, or completely spatially
random, over the domain $[0, 1] \times [0, 1]$. For a larger domain, they are not uniform,
for the two square windows w1 and w2 created by

```
w1 <- st_bbox(c(xmin = 0, ymin = 0, xmax = 1, ymax = 1)) |>
        st_as_sfc()
w2 <- st_sfc(st_point(c(1, 0.5))) |> st_buffer(1.2)
```

this is shown in Figure 11.1.

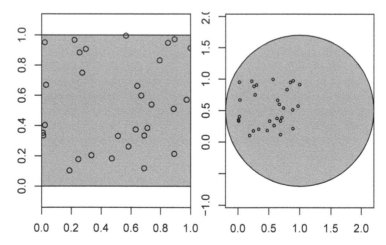

Figure 11.1: Depending on the observation window (grey), the same point
pattern can appear completely spatially random (left), or clustered (right)

Point patterns in **spatstat** are objects of class **ppp** that contain points and an
observation window (an object of class **owin**). We can create a **ppp** from points by

```
library(spatstat) |> suppressPackageStartupMessages()
as.ppp(xy)
# Planar point pattern: 30 points
# window: rectangle = [0.009, 0.999] x [0.103, 0.996] units
```

where we see that the bounding box of the points is used as observation window when no window is specified. If we add a polygonal geometry as the first feature of the dataset, then this is used as observation window:

```
(pp1 <- c(w1, st_geometry(xy)) |> as.ppp())
# Planar point pattern: 30 points
# window: polygonal boundary
# enclosing rectangle: [0, 1] x [0, 1] units
c1 <- st_buffer(st_centroid(w2), 1.2)
(pp2 <- c(c1, st_geometry(xy)) |> as.ppp())
# Planar point pattern: 30 points
# window: polygonal boundary
# enclosing rectangle: [-0.2, 2.2] x [-0.7, 1.7] units
```

To test for homogeneity, one could carry out a quadrat count, using an appropriate quadrat layout (a 3×3 layout is shown in Figure 11.2)

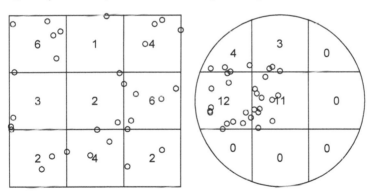

Figure 11.2: 3×3 quadrat counts for the two point patterns

and carry out a χ^2 test on these counts:

```
quadrat.test(pp1, nx=3, ny=3)
# Warning: Some expected counts are small; chi^2 approximation
# may be inaccurate
#
#   Chi-squared test of CSR using quadrat counts
#
# data:  pp1
# X2 = 8, df = 8, p-value = 0.9
```

```
# alternative hypothesis: two.sided
#
# Quadrats: 9 tiles (irregular windows)
quadrat.test(pp2, nx=3, ny=3)
# Warning: Some expected counts are small; chi^2 approximation
# may be inaccurate
#
#   Chi-squared test of CSR using quadrat counts
#
# data:  pp2
# X2 = 43, df = 8, p-value = 2e-06
# alternative hypothesis: two.sided
#
# Quadrats: 9 tiles (irregular windows)
```

which indicates that for the second case we have an indication that this is not a CSR (completely spatially random) pattern. As indicated by the warning, we should take the p-values with a large grain of salt because we have too small expected counts.

Kernel densities can be computed using **density**, where kernel shape and bandwidth can be controlled. Here, cross-validation is used by function **bw.diggle** to specify the bandwidth parameter **sigma**; plots are shown in Figure 11.3.

```
den1 <- density(pp1, sigma = bw.diggle)
den2 <- density(pp2, sigma = bw.diggle)
```

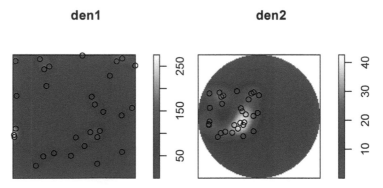

Figure 11.3: Kernel densities for both point patterns

The density maps created this way are obviously raster images, and we can convert them into stars object by

```
library(stars)
# Loading required package: abind
s1 <- st_as_stars(den1)
```

```
(s2 <- st_as_stars(den2))
# stars object with 2 dimensions and 1 attribute
# attribute(s):
#         Min.  1st Qu. Median Mean 3rd Qu. Max. NA's
# v  1.03e-14 0.000153  0.304 6.77    13.1 42.7 3492
# dimension(s):
#   from  to offset    delta x/y
# x    1 128   -0.2 0.01875 [x]
# y    1 128   -0.7 0.01875 [y]
```

and we can verify that the area under the density surface is similar to the sample
size (30), by

```
s1$a <- st_area(s1) |> suppressMessages()
s2$a <- st_area(s2) |> suppressMessages()
with(s1, sum(v * a, na.rm = TRUE))
# [1] 29
with(s2, sum(v * a, na.rm = TRUE))
# [1] 30.7
```

More exciting applications involve modelling the density surface as a function of
external variables. Suppose we want to model the density of **pp2** as a Poisson point
process (meaning that points do not interact with each other), where the intensity is
a function of distance to the centre of the "cluster", and these distance are available
in a **stars** object:

```
pt <- st_sfc(st_point(c(0.5, 0.5)))
st_as_sf(s2, as_points = TRUE, na.rm = FALSE) |>
  st_distance(pt) -> s2$dist
```

we can then model the densities using **ppm**, where the *name* of the point pattern
object is used on the left-hand side of the **formula**:

```
(m <- ppm(pp2 ~ dist, data = list(dist = as.im(s2["dist"]))))
# Nonstationary Poisson process
#
# Log intensity:  ~dist
#
# Fitted trend coefficients:
# (Intercept)        dist
#        4.54       -4.25
#
#
#              Estimate S.E. CI95.lo CI95.hi Ztest  Zval
# (Intercept)      4.54 0.341    3.87    5.21   *** 13.32
# dist            -4.25 0.701   -5.62   -2.88   *** -6.06
```

The returned object is of class **ppm**, and can be plotted: Figure 11.4 shows the predicted surface. The prediction standard error can also be plotted.

Fitted trend

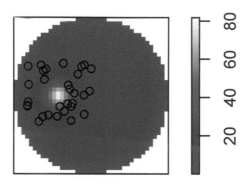

Figure 11.4: Predicted densities of a ppm model

The model also has a **predict** method, which returns an **im** object that can be converted into a **stars** object by

```
predict(m, covariates = list(dist = as.im(s2["dist"]))) |>
    st_as_stars()
# stars object with 2 dimensions and 1 attribute
# attribute(s):
#       Min. 1st Qu. Median Mean 3rd Qu. Max. NA's
# v  0.0694   0.527   2.12 6.62     7.3 89.9 3492
# dimension(s):
#   from  to offset   delta x/y
# x    1 128   -0.2 0.01875 [x]
# y    1 128   -0.7 0.01875 [y]
```

11.2   Coordinate reference systems

All routines in **spatstat** are layed out for two-dimensional data with Cartesian coordinates. If we try to convert an object with ellipsoidal coordinates, we get an error:

```
system.file("gpkg/nc.gpkg", package = "sf") |>
    read_sf() |>
    st_geometry() |>
```

```
     st_centroid() |>
     as.ppp()
# Error: Only projected coordinates may be converted to
# spatstat class objects
```

When converting to a **spatstat** data structure we loose the coordinate reference system we started with. It can be set back to **sf** or **stars** objects by using **st_set_crs**.

11.3 Marked point patterns, points on linear networks

A few more data types can be converted to and from **spatstat**. Marked point patterns are point patterns that have a "mark", which is either a categorical label or a numeric label for each point. A dataset available in **spatstat** with marks is the **longleaf** pines dataset, containing diameter at breast height as a numeric mark:

```
longleaf
# Marked planar point pattern: 584 points
# marks are numeric, of storage type  'double'
# window: rectangle = [0, 200] x [0, 200] metres
ll <- st_as_sf(longleaf)
print(ll, n = 3)
# Simple feature collection with 585 features and 2 fields
# Geometry type: GEOMETRY
# Dimension:     XY
# Bounding box:  xmin: 0 ymin: 0 xmax: 200 ymax: 200
# CRS:           NA
# First 3 features:
#     spatstat.geom..marks.x.  label
# NA                       NA window
# 1                      32.9  point
# 2                      53.5  point
#                               geom
# NA POLYGON ((0 0, 200 0, 200 2...
# 1                POINT (200 8.8)
# 2                 POINT (199 10)
```

Values can be converted back to **ppp** with

```
as.ppp(ll)
# Warning in as.ppp.sf(ll): only first attribute column is used
# for marks
# Marked planar point pattern: 584 points
# marks are numeric, of storage type  'double'
```

```
# window: polygonal boundary
# enclosing rectangle: [0, 200] x [0, 200] units
```

Line segments, in **spatstat** objects of class **psp** can be converted back and forth
to simple feature with **LINESTRING** geometries following a **POLYGON** feature with the
observation window, as in

```
print(st_as_sf(copper$SouthLines), n = 5)
# Simple feature collection with 91 features and 1 field
# Geometry type: GEOMETRY
# Dimension:     XY
# Bounding box:  xmin: -0.335 ymin: 0.19 xmax: 35 ymax: 158
# CRS:           NA
# First 5 features:
#     label                         geom
# 1   window POLYGON ((-0.335 0.19, 35 0...
# 2  segment LINESTRING (3.36 0.19, 10.4...
# 3  segment LINESTRING (12.5 0.263, 11....
# 4  segment LINESTRING (11.2 0.197, -0....
# 5  segment LINESTRING (6.35 12.8, 16.5...
```

Finally, point patterns on linear networks, in **spatstat** represented by **lpp** objects,
can be converted to **sf** by

```
print(st_as_sf(chicago), n = 5)
# Simple feature collection with 620 features and 4 fields
# Geometry type: GEOMETRY
# Dimension:     XY
# Bounding box:  xmin: 0.389 ymin: 153 xmax: 1280 ymax: 1280
# CRS:           NA
# First 5 features:
#     label seg tp marks                         geom
# 1   window  NA NA  <NA> POLYGON ((0.389 153, 1282 1...
# 2  segment  NA NA  <NA> LINESTRING (0.389 1254, 110...
# 3  segment  NA NA  <NA> LINESTRING (110 1252, 111 1...
# 4  segment  NA NA  <NA> LINESTRING (110 1252, 198 1...
# 5  segment  NA NA  <NA> LINESTRING (198 1277, 198 1...
```

where we only see the first five features; the points are also in this object, as variable
label indicates

```
table(st_as_sf(chicago)$label)
#
#   point segment  window
#     116     503       1
```

Potential information about network *structure*, how LINESTRING geometries are connected, is not present in the **sf** object. Package **sfnetworks** (van der Meer et al. 2022) would be a candidate package to hold such information in R or to pass on network data imported from OpenStreetMaps to **spatstat**.

11.4   Spatial sampling and simulating a point process

Package **sf** contains an st_sample method that samples points from MULTIPOINT, linear or polygonal geometries, using different spatial sampling strategies. It natively supports strategies "random", "hexagonal", "Fibonacci" (Section 11.5), and "regular", where "regular" refers to sampling on a square regular grid and "hexagonal" essentially gives a triangular grid. For type "random", it can return exactly the number of requested points, for other types this is approximate.

st_sample also interfaces point process simulation functions of **spatstat**, when other values for sampling type are chosen. For instance the **spatstat** function rThomas is invoked when setting **type** ▪ Thomas (Figure 11.5):

```
kappa <- 30 / st_area(w2) # intensity
th <- st_sample(w2, kappa = kappa, mu = 3, scale = 0.05,
    type = "Thomas")
nrow(th)
# [1] 90
```

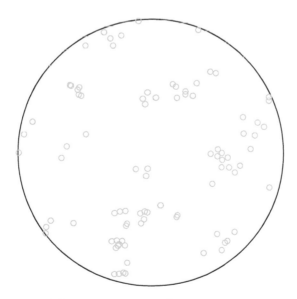

Figure 11.5: Thomas process with mu = 3 and scale = 0.05

The help function obtained by ?rThomas details the meaning of the parameters kappa, mu, and scale. Simulating point processes means that the intensity is given, not the sample size. The sample size within the observation window obtained this way is a random variable.

11.5 Simulating points on the sphere

Another spatial random sampling type supported by sf natively (in st_sample) is simulation of random points on the sphere. An example of this is shown in Figure 11.6, where points were constrained to those in oceans. Points approximately regularly distributed over a sphere are obtained by st_sample with type = "Fibonacci" (González 2010).

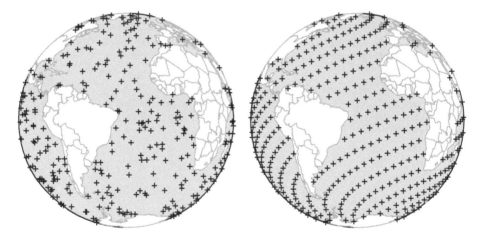

Figure 11.6: Points sampled on the globe over the oceans: randomly (left) and approximately regular (Fibonacci; right), shown in an orthographic projection

11.6 Exercises

1. After loading **spatstat**, recreate the plot obtained by plot(longleaf) by using **ggplot2** and geom_sf(), and by sf::plot().
2. Convert the sample locations of the NO_2 data used in Chapter 12 to a ppp object, with a proper window.
3. Compute and plot the density of the NO_2 dataset, import the density as a **stars** object, and compute the volume under the surface.

12

Spatial Interpolation

Spatial interpolation is the activity of estimating values of spatially continuous variables (fields) for spatial locations where they have not been observed, based on observations. The statistical methodology for spatial interpolation, called geostatistics, is concerned with the modelling, prediction, and simulation of spatially continuous phenomena. The typical problem is a missing value problem: we observe a property of a phenomenon $Z(s)$ at a limited number of sample locations $s_i, i = 1, ..., n$, and are interested in the property value at all locations s_0 covering an area of interest, so we have to predict it for unobserved locations. This is also called *kriging*, or Gaussian Process prediction. In case $Z(s)$ contains a white noise component ϵ, as in $Z(s) = S(s) + \epsilon$ (possibly reflecting measurement error) an alternative but similar goal is to predict or simulate $S(s)$ rather than $Z(s)$, which may be called *spatial filtering* or *smoothing*.

In this chapter we will show simple approaches for handling geostatistical data, demonstrate simple interpolation methods, and explore modelling spatial correlation, spatial prediction and simulation. Chapter 13 focuses on more complex multivariate and spatiotemporal geostatistical models. We will use package **gstat** (Pebesma and Graeler 2022; Pebesma 2004), which offers a fairly wide palette of models and options for non-Bayesian geostatistical analysis. Bayesian methods with R implementations are found in Diggle, Tawn, and Moyeed (1998), Diggle and Ribeiro Jr. (2007), Blangiardo and Cameletti (2015), and Wikle, Zammit-Mangion, and Cressie (2019). An overview and comparison of methods for large datasets is given in Heaton et al. (2018).

12.1 A first dataset

We can read station mean NO_2 values, a dataset that is prepared in Chapter 13, by loading it from package **gstat** using

```
library(tidyverse) |> suppressPackageStartupMessages()
no2 <- read_csv(system.file("external/no2.csv",
    package = "gstat"), show_col_types = FALSE)
```

and convert it into an **sf** object with an appropriate UTM projection using

```
library(sf)
# Linking to GEOS 3.11.1, GDAL 3.6.2, PROJ 9.1.1; sf_use_s2()
# is TRUE
crs <- st_crs("EPSG:32632")
st_as_sf(no2, crs = "OGC:CRS84", coords =
    c("station_longitude_deg", "station_latitude_deg")) |>
    st_transform(crs) -> no2.sf
```

Next, we can load country boundaries and plot these data using **ggplot**, shown in
Figure 12.1.

```
read_sf("data/de_nuts1.gpkg") |> st_transform(crs) -> de
```

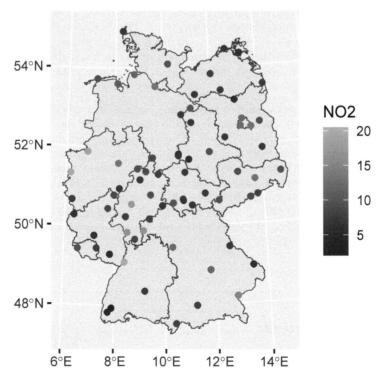

Figure 12.1: Mean NO_2 concentrations in air for rural background stations in
Germany, in 2017

If we want to interpolate, we first need to decide where. This is typically done on a
regular grid covering the area of interest. Starting with the country outline in object
de we can create a regular 10 km × 10 km grid over Germany by

```
library(stars) |> suppressPackageStartupMessages()
st_bbox(de) |>
  st_as_stars(dx = 10000) |>
  st_crop(de) -> grd
grd
# stars object with 2 dimensions and 1 attribute
# attribute(s):
#            Min. 1st Qu. Median Mean 3rd Qu. Max. NA's
# values     0      0      0     0     0      0  2076
# dimension(s):
#    from to  offset  delta           refsys x/y
# x     1 65  280741   10000 WGS 84 / UTM z... [x]
# y     1 87 6101239  -10000 WGS 84 / UTM z... [y]
```

Here, we chose grid cells not too fine, so that we still see them in plots.

Perhaps the simplest interpolation method is inverse distance weighted interpolation, which is a weighted average, using weights inverse proportional to distances from the interpolation location:

$$\hat{z}(s_0) - \frac{\sum_{i=1}^{n} w_i z(s_i)}{\sum_{i=1}^{n} w_i}$$

with $w_i = |s_0 - s_i|^{-p}$, and the inverse distance power p typically taken as 2, or optimised using cross-validation. We can compute inverse distance interpolated values using **gstat::idw**,

```
library(gstat)
i <- idw(NO2~1, no2.sf, grd)
# [inverse distance weighted interpolation]
```

and plot them in Figure 12.2.

12.2 Sample variogram

In order to make spatial predictions using geostatistical methods, we first need to identify a model for the mean and for the spatial correlation. In the simplest model, $Z(s) = m + e(s)$, the mean is an unknown constant m, and in this case the spatial correlation can be modelled using the variogram, $\gamma(h) = 0.5E(Z(s) - Z(s + h))^2$. For processes with a finite variance $C(0)$, the variogram is related to the covariogram or covariance function through $\gamma(h) = C(0) - C(h)$.

Spatial Interpolation

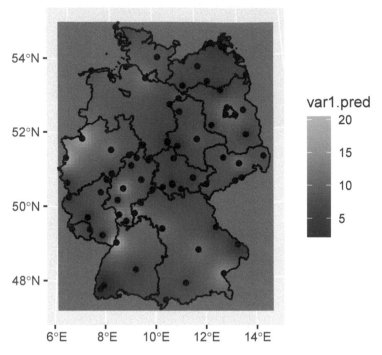

Figure 12.2: Inverse distance weighted interpolated values for NO_2 over Germany

The sample variogram is obtained by computing estimates of $\gamma(h)$ for distance intervals, $h_i = [h_{i,0}, h_{i,1}]$:

$$\hat{\gamma}(h_i) = \frac{1}{2N(h_i)} \sum_{j=1}^{N(h_i)} (z(s_i) - z(s_i + h'))^2, \quad h_{i,0} \le h' < h_{i,1} \qquad (12.1)$$

with $N(h_i)$ the number of sample pairs available for distance interval h_i. Function `gstat::variogram` computes sample variograms,

```
v <- variogram(NO2~1, no2.sf)
```

and the result of plotting this is shown in Figure 12.3.

Function **variogram** chooses default for maximum distance (**cutoff**: one-third of the length of the bounding box diagonal) and (constant) interval widths (**width**: cutoff divided by 15). These defaults can be changed by

```
v0 <- variogram(NO2~1, no2.sf, cutoff = 100000, width = 10000)
```

shown in Figure 12.4.

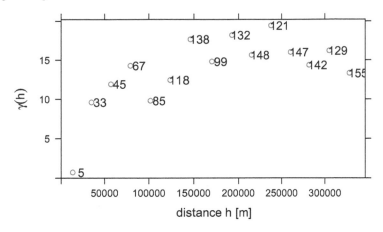

Figure 12.3: Sample variogram plot

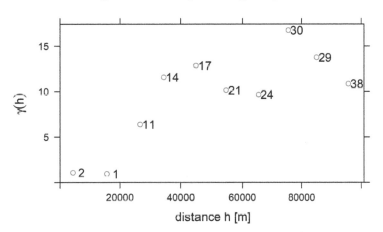

Figure 12.4: Sample variogram plot with adjusted cutoff and lag width

Note that the formula NO2~1 is used to select the variable of interest from the data file (NO2), and to specify the mean model: ~1 specifies an intercept-only (unknown, constant mean) model.

12.3 Fitting variogram models

In order to progress towards spatial predictions, we need a variogram *model* $\gamma(h)$ for (potentially) all distances h, rather than the set of estimates derived above. If we would connect these estimates with straight lines, or assume they reflect constant values over their respective distance intervals, it would lead to statistical models with non-positive definite covariance matrices, which would block using them in prediction.

To avoid this, we fit parametric models $\gamma(h)$ to the estimates $\hat{\gamma}(h_i)$, where we take h_i as the mean value of all the h' values involved in estimating $\hat{\gamma}(h_i)$. We can fit for instance a model with an exponential variogram by

```
v.m <- fit.variogram(v, vgm(1, "Exp", 50000, 1))
```

shown by the solid line in Figure 12.5.

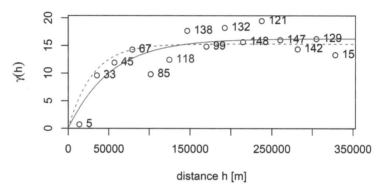

Figure 12.5: Sample variogram (circles) with models fitted using weighted least squares (solid line) and maximum likelihood estimation (dashed line)

The fitting for the drawn line was done by weighted least squares, minimising

$$\sum_{i=1}^{n} w_i (\gamma(h_i) - \hat{\gamma}(h_i))^2, \tag{12.2}$$

with weights w_i by default equal to $N(h_i)/h^2$. Other weight options are available through argument `fit.method`.

As an alternative to weighted least squares fitting, one can use maximum likelihood (ML) or restricted maximum likelihood parameter estimation (Kitanidis and Lane 1985), which for this case leads to a relatively similar fitted model, shown as the dashed line in Figure 12.5. An advantage of ML-type approaches is that they do not require choosing distance intervals h_i in Equation 12.1 or weights w_i in Equation 12.2. Disadvantages are that they lean on stronger assumptions of multivariate normally distributed data, and for larger datasets require iteratively solving linear systems of size equal to the number of observations; Heaton et al. (2018) compare approaches dedicated to fitting models to large datasets.

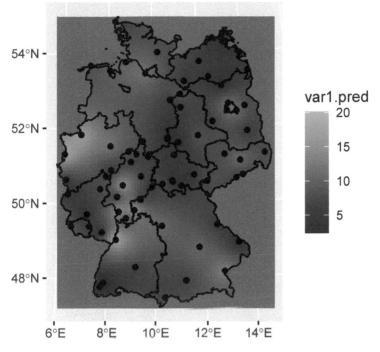

Figure 12.6: Kriged NO$_2$ concentrations over Germany

12.4   Kriging interpolation

Typically, when we interpolate a variable, we do that on points on a regular grid covering the target area. We first create a **stars** object with a raster covering the target area, and **NAs** outside it.

Kriging involves the prediction of $Z(s_0)$ at arbitrary locations s_0. We can krige NO$_2$ by using **gstat::krige**, with the model for the trend, the data, the prediction grid, and the variogram model as arguments (Figure 12.6) by:

```
k <- krige(NO2~1, no2.sf, grd, v.m)
# [using ordinary kriging]
```

12.5   Areal means: block kriging

Computing areal means can be done in several ways. The simplest is to take the average of point samples falling inside the target polygons:

```
a <- aggregate(no2.sf["NO2"], by = de, FUN = mean)
```

A more complicated way is to use *block kriging* (Journel and Huijbregts 1978), which uses *all* the data to estimate the mean of the variable over the target areas. With **krige**, this can be done by giving the target areas (polygons) as the **newdata** argument:

```
b <- krige(NO2~1, no2.sf, de, v.m)
# [using ordinary kriging]
```

we can now merge the two maps into a single object to create a single plot (Figure 12.7):

```
b$sample <- a$NO2
b$kriging <- b$var1.pred
```

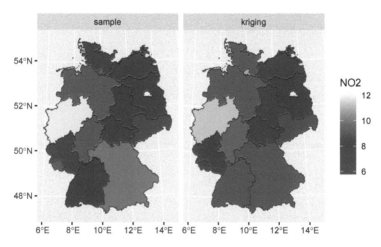

Figure 12.7: Aggregated NO$_2$ values from simple averaging (left) and block kriging (right)

We see that the signal is similar, but that the sample means from simple averaging are more variable than the block kriging values; this may be due to the smoothing effect of kriging: data points outside the aggregation area receive weight, too.

To compare the standard errors of means, for the sample mean we can get a rough guess of the standard error by $\sqrt{(\sigma^2/n)}$:

```
SE <- function(x) sqrt(var(x)/length(x))
a <- aggregate(no2.sf["NO2"], de, SE)
```

which would have been the actual estimate in design-based inference (Section 10.4) if the sample were obtained by spatially random sampling. The block kriging variance is the model-based estimate and is a by-product of kriging. We can compare the two in Figure 12.8 where we see that the simple averaging approach gives more variability and mostly larger values for prediction errors of areal means, compared to block kriging.

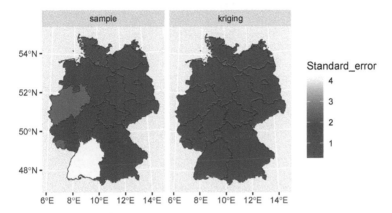

Figure 12.8: Standard errors for mean NO_2 values obtained by simple averaging (left) and block kriging (right)

12.6   Conditional simulation

In case one or more conditional realisation of the field $Z(s)$ are needed rather than their conditional mean, we can obtain this by *conditional simulation*. A reason for wanting this may be the need to estimate areal mean values of $g(Z(s))$ with $g(\cdot)$ a non-linear function; a simple example is the areal fraction where $Z(s)$ exceeds a threshold.

The default approach used by **gstat** is to use the sequential simulation algorithm for this. This is a simple algorithm that randomly steps through the prediction locations and at each location:

- carries out a kriging prediction
- draws a random variable from the normal distribution with mean and variance equal to the kriging variance
- adds this value to the conditioning dataset
- finds a new random simulation location

until all locations have been visited.

This is carried out by **gstat::krige** when **nsim** is set to a positive value:

```
set.seed(13341)
(s <- krige(NO2~1, no2.sf, grd, v.m, nmax = 30, nsim = 6))
# drawing 6 GLS realisations of beta...
# [using conditional Gaussian simulation]
# stars object with 3 dimensions and 1 attribute
# attribute(s):
#         Min. 1st Qu. Median Mean 3rd Qu. Max.  NA's
# var1   -5.7    6.12   8.68 8.88    11.5 23.9 12456
# dimension(s):
#          from to  offset  delta           refsys      values
# x           1 65  280741  10000 WGS 84 / UTM z...        NULL
# y           1 87 6101239 -10000 WGS 84 / UTM z...        NULL
# sample      1  6      NA     NA              NA sim1,...,sim6
#             x/y
# x           [x]
# y           [y]
# sample
```

where **set.seed()** was called here to allow reproducibility.

It is usually needed to constrain the (maximum) number of nearest neighbours to include in kriging estimation by setting **nmax** because the dataset grows each step, leading otherwise quickly to very long computing times and large memory requirements. Resulting conditional simulations are shown in (Figure 12.9).

Alternative methods for conditional simulation have recently been added to **gstat**, and include **krigeSimCE** implementing the circular embedding method (Davies and Bryant 2013), and **krigeSTSimTB** implementing the turning bands method (Schlather 2011). These are of particular of interest for larger datasets or conditional simulations of spatiotemporal data.

12.7 Trend models

Kriging and conditional simulation, as used so far in this chapter, assume that all spatial variability is a random process, characterised by a spatial covariance model. In case we have other variables that are meaningfully correlated with the target variable, we can use them in a linear regression model for the trend,

$$Z(s) = \sum_{j=0}^{p} \beta_j X_j(s) + e(s)$$

with $X_0(s) = 1$ and β_0 an intercept, but with the other β_j regression coefficients. Adding variables typically reduces both the spatial correlation in the residual $e(s)$, as well as its variance, and leads to more accurate predictions and more similar

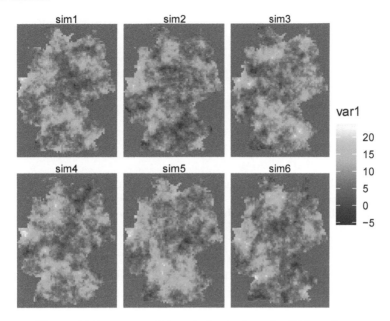

Figure 12.9: Six conditional simulations for NO$_2$ values

conditional simulations. As an example, we will use population density to partly explain variation in NO$_2$.

12.7.1 A population grid

As a potential predictor for NO$_2$ in the air, we use population density. NO$_2$ is mostly caused by traffic, and traffic is more intense in densely populated areas. Population density is obtained from the 2011 census and is downloaded as a csv file with the number of inhabitants per 100 m × 100 m grid cell. We can aggregate these data to the target grid cells by summing the inhabitants:

```
v <- vroom::vroom("aq/pop/Zensus_Bevoelkerung_100m-Gitter.csv")
v |> filter(Einwohner > 0) |>
    select(-Gitter_ID_100m) |>
    st_as_sf(coords = c("x_mp_100m", "y_mp_100m"), crs = 3035) |>
    st_transform(st_crs(grd)) -> b
a <- aggregate(b, st_as_sf(grd, na.rm = FALSE), sum)
```

Now we have the population counts per grid cell in **a**. To get to population density, we need to find the area of each cell; for cells crossing the country border, this will be less than 10 × 10 km:

```
grd$ID <- 1:prod(dim(grd)) # to identify grid cells
ii <- st_intersects(grd["ID"],
  st_cast(st_union(de), "MULTILINESTRING"), as_points = FALSE)
grd_sf <- st_as_sf(grd["ID"], na.rm = FALSE)[lengths(ii) > 0,]
st_agr(grd_sf) = "identity"
iii <- st_intersection(grd_sf, st_union(de))
grd$area <- st_area(grd)[[1]] +
    units::set_units(grd$values, m^2)
grd$area[iii$ID] <- st_area(iii)
```

Instead of doing the two-stage procedure above, first finding cells that have a border crossing it then computing its area, we could also directly use **st_intersection** on all cells, but that takes considerably longer. From the counts and areas we can compute densities (Figure 12.10) and verify totals

```
grd$pop_dens <- a$Einwohner / grd$area
sum(grd$pop_dens * grd$area, na.rm = TRUE) # verify
# 80323301 [1]
sum(b$Einwohner)
# [1] 80324282
```

which indicates strong agreement. Using **st_interpolate_aw** would have given an exact match.

We need to divide the number of inhabitants by the number of 100 m × 100 m grid cells contributing to it, in order to convert population counts into population density.

To obtain population density values at monitoring network stations, we can use **st_extract**:

```
grd |>
  select("pop_dens") |>
  st_extract(no2.sf) |>
  pull("pop_dens") |>
  mutate(no2.sf, pop_dens = _) -> no2.sf
```

We can then investigate the linear relationship between NO$_2$ and population density at monitoring station locations:

```
summary(lm(NO2~sqrt(pop_dens), no2.sf))
#
# Call:
# lm(formula = NO2 ~ sqrt(pop_dens), data = no2.sf)
#
# Residuals:
#     Min     1Q Median     3Q    Max
# -7.990 -2.052 -0.505  1.610  8.095
```

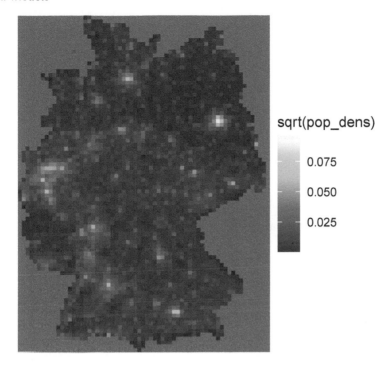

Figure 12.10: Population density for 100 m × 100 m grid cells

```
#
# Coefficients:
#                 Estimate Std. Error t value Pr(>|t|)
# (Intercept)       4.537      0.685     6.62  5.5e-09 ***
# sqrt(pop_dens)  326.154     49.366     6.61  5.8e-09 ***
# ---
# Signif. codes:  0 '***' 0.001 '**' 0.01 '*' 0.05 '.' 0.1 ' ' 1
#
# Residual standard error: 3.13 on 72 degrees of freedom
# Multiple R-squared:  0.377,   Adjusted R-squared:  0.369
# F-statistic: 43.7 on 1 and 72 DF,  p-value: 5.82e-09
```

for which the corresponding scatterplot is shown in Figure 12.11.

Prediction under this new model involves first modelling a residual variogram

```
no2.sf <- no2.sf[!is.na(no2.sf$pop_dens),]
vr <- variogram(NO2~sqrt(pop_dens), no2.sf)
vr.m <- fit.variogram(vr, vgm(1, "Exp", 50000, 1))
```

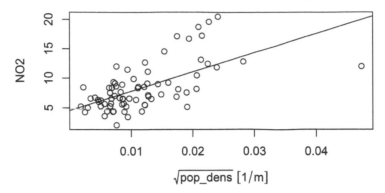

Figure 12.11: Scatter plot of 2017 annual mean NO_2 concentration against population density, for rural background air quality stations

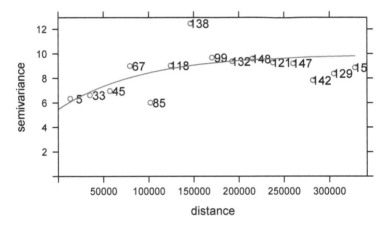

Figure 12.12: Residual variogram after subtracting population density trend

which is shown in Figure 12.12. Subsequently, kriging prediction (Figure 12.13) is done by

```
kr <- krige(NO2 ~ sqrt(pop_dens), no2.sf,
            grd["pop_dens"], vr.m)
# [using universal kriging]
```

where, critically, the **pop_dens** values are now available for prediction locations in the **newdata** object **grd**.

Compared to (ordinary) kriging We see some clear differences: the map using population density in the trend follows the extremes of the population density rather than those of the measurement stations, and has a value range that extends that of ordinary kriging. It should be taken with a large grain of salt however, since the stations used were filtered for the category "rural background", indicating that they only

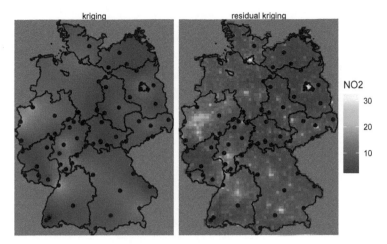

Figure 12.13: Kriging NO_2 values using population density as a trend variable

represent conditions of lower populations density. The scatter-plot of Figure 12.11 reveals that the the population density at the locations of stations is much more limited than that in the population density map, and hence the right-hand side map is based on strongly extrapolating the relationship shown in Figure 12.11.

12.8 Exercises

1. Create a plot like the one in Figure 12.13 that has the inverse distance interpolated map of Figure 12.2 added on the left side.
2. Create a scatter-plot of the map values of the idw and kriging map, and a scatter-plot of map values of idw and residual kriging.
3. Carry out a *block kriging*, predicting block averages for blocks centred over grid cells, by setting the **block** argument in **krige()**, and do this for block sizes of 10 km (grid cell size), 50 km, and 200 km. Compare the resulting maps of estimates for these three block sizes with those obtained by point kriging, and do the same thing for all associated kriging standard errors.
4. Based on the residual kriging results obtained above, compute maps of the lower and upper boundary of a 95% confidence interval, when assuming that the kriging error is normally distributed, and show them in a plot with a single (joint) legend.
5. Compute and show the map with the probabilities that NO_2 point values exceed the level of 15 ppm, assuming normally distributed kriging errors.

13

Multivariate and Spatiotemporal Geostatistics

Building on the simple interpolation methods presented in Chapter 12, this chapter continues with multivariate geostatistics and spatiotemporal geostatistics. The topic of multivariate geostatistics, more extensively illustrated in Bivand, Pebesma, and Gómez-Rubio (2013), is briefly introduced. Spatiotemporal geostatistics is illustrated with a worked out case study for spatiotemporal interpolation, using NO_2 air quality data, and population density as covariate.

13.1 Preparing the air quality dataset

The dataset we work with is an air quality dataset obtained from the European Environmental Agency (EEA). European member states report air quality measurements to this agency. So-called *validated* data are quality controlled by member states, and are reported on a yearly basis. They form the basis for policy compliancy evaluations and (counter) measures.

The EEA's air quality e-reporting website gives access to the data reported by European member states. We decided to download hourly (time series) data, which is the data primarily measured. A web form helps convert simple selection criteria into an http **GET** request. The URL[1] was created to select all validated (**Source=E1a**) NO_2 (**Pollutant=8**) data for 2017 (**Year_from, Year_to**) from Germany (**CountryCode=DE**). It returns a text file with a set of URLs to CSV files, each containing the hourly values for the whole period for a single measurement station. These files were downloaded and converted to the right encoding using the **dos2unix** command line utility.

In the following, we will read all the files into a list,

```
files <- list.files("aq", pattern = "*.csv", full.names = TRUE)
r <- lapply(files[-1], function(f) read.csv(f))
```

then convert the time variable into a **POSIXct** variable, and put them in time order by

---

[1] https://fme.discomap.eea.europa.eu/fmedatastreaming/AirQualityDownload/AQData_Extract.fmw?CountryCode=DE&CityName=&Pollutant=8&Year_from=2017&Year_to=2017&Station=&Samplingpoint=&Source=E1a&Output=TEXT&UpdateDate=

```
Sys.setenv(TZ = "UTC") # don't use local time zone
r <- lapply(r, function(f) {
        f$t = as.POSIXct(f$DatetimeBegin)
        f[order(f$t), ]
    }
)
```

We remove smaller sub-datasets, which for this dataset have no hourly data:

```
r <- r[sapply(r, nrow) > 1000]
names(r) <- sapply(r,
                function(f) unique(f$AirQualityStationEoICode))
length(r) == length(unique(names(r)))
# [1] TRUE
```

and then merge all files using **xts::cbind**, so that records are combined based on matching times:

```
library(xts) |> suppressPackageStartupMessages()
r <- lapply(r, function(f) xts(f$Concentration, f$t))
aq <- do.call(cbind, r)
```

A usual further selection for this dataset is to select stations for which 75% of the hourly values measured are valid, i.e., drop those with more than 25% missing hourly values. Knowing that **mean(is.na(x))** gives the *fraction* of missing values in a vector **x**, we can apply this function to the columns (stations):

```
sel <- apply(aq, 2, function(x) mean(is.na(x)) < 0.25)
aqsel <- aq[, sel]
```

Next, the station metadata was read and filtered for rural background stations in Germany (**"DE"**) by

```
library(tidyverse) |> suppressPackageStartupMessages()
read.csv("aq/AirBase_v8_stations.csv", sep = "\t") |>
    as_tibble() |>
    filter(country_iso_code == "DE",
            station_type_of_area == "rural",
            type_of_station == "Background") -> a2
```

These stations contain coordinates, and an **sf** object with (static) station metadata is created by

```
library(sf) |> suppressPackageStartupMessages()
a2.sf <- st_as_sf(a2, crs = 'OGC:CRS84',
  coords = c("station_longitude_deg", "station_latitude_deg"))
```

We now subset the air quality measurements to include only stations that are of type rural background, which we saved in **a2**:

```
sel <- colnames(aqsel) %in% a2$station_european_code
aqsel <- aqsel[, sel]
dim(aqsel)
# [1] 8760   74
```

We can compute station means and join these to station locations by

```
tb <- tibble(NO2 = apply(aqsel, 2, mean, na.rm = TRUE),
             station_european_code = colnames(aqsel))
crs <- st_crs('EPSG:32632')
right_join(a2.sf, tb) |> st_transform(crs) -> no2.sf
read_sf("data/de_nuts1.gpkg") |> st_transform(crs) -> de
```

Station mean NO_2 concentrations, along with country borders, are shown in in Figure 12.1.

13.2 Multivariable geostatistics

Multivariable geostatics involves the *joint* modelling, prediction, and simulation of multiple variables,

$$Z_1(s) = X_1\beta_1 + e_1(s)$$

$$\dots$$

$$Z_n(s) = X_n\beta_n + e_n(s).$$

In addition to having observations, trend models, and variograms for each variable, the *cross*-variogram for each pair of residual variables, describing the covariance of $e_i(s), e_j(s+h)$, is required. If this cross-covariance is non-zero, knowledge of $e_j(s+h)$ may help predict (or simulate) $e_i(s)$. This is especially true if $Z_j(s)$ is more densely sample than $Z_i(s)$. Prediction and simulation under this model are called cokriging and cosimulation. Examples using gstat are found when running the demo scripts

```
library(gstat)
demo(cokriging)
demo(cosimulation)
```

and are further illustrated and discussed in Bivand, Pebesma, and Gómez-Rubio (2013).

In case the different variables considered are observed at the same set of locations, for instance different air quality parameters, then the statistical *gain* of using cokriging as opposed to direct (univariable) kriging is often modest, when not negligible. A gain may however be that the prediction is truly multivariable: in addition to the prediction vector $Z(\hat{s}_0) = (\hat{Z}_1(s_0), ..., \hat{Z}_n(s_0))$, we get the full covariance matrix of the prediction error (Ver Hoef and Cressie 1993). Using these prediction error covariances, for any linear combination of $\hat{Z}(s_0)$, such as $\hat{Z}_2(s_0) - \hat{Z}_1(s_0)$, we can get the standard error of that combination.

Although sets of direct and cross-variograms can be computed and fitted automatically, multivariable geostatistical modelling becomes quickly hard to manage when the number of variables gets large, because the number of direct and cross-variograms required is $n(n + 1)/2$.

In case different variables refer to the same variable taken at different time steps, one could use a multivariable (cokriging) prediction approach, but this would not allow for interpolation between two time steps. For this, and for handling the case of having data observed at many time instances, one can also model its variation as a function of continuous space *and* time, as of $Z(s, t)$, which we will do in the next section.

13.3   Spatiotemporal geostatistics

Spatiotemporal geostatistical processes are modelled as variables having a value everywhere in space and time, $Z(s, t)$, with s and t the continuously indexed space and time index. Given observations $Z(s_i, t_j)$ and a variogram (covariance) model $\gamma(s, t)$ we can predict $Z(s_0, t_0)$ at arbitrary space/time locations (s_0, t_0) using standard Gaussian process theory.

Several books have been written recently about modern approaches to handling and modelling spatiotemporal geostatistical data, including Wikle, Zammit-Mangion, and Cressie (2019) and Blangiardo and Cameletti (2015). Here, we will use Gräler, Pebesma, and Heuvelink (2016) and give some simple examples using the dataset also used for the previous chapter.

13.3.1   A spatiotemporal variogram model

Starting with the spatiotemporal matrix of NO$_2$ data in `aq` constructed at the beginning of this chapter, we selected complete records taken at rural background stations into `aqsel`. We can select the spatial locations for these 74 stations by

```
sfc <- st_geometry(a2.sf)[match(colnames(aqsel),
                         a2.sf$station_european_code)] |>
   st_transform(crs)
```

and finally build a **stars** vector data cube with time and station as dimensions:

```
library(stars)
# Loading required package: abind
st_as_stars(NO2 = as.matrix(aqsel)) |>
    st_set_dimensions(names = c("time", "station")) |>
    st_set_dimensions("time", index(aqsel)) |>
    st_set_dimensions("station", sfc) -> no2.st
no2.st
# stars object with 2 dimensions and 1 attribute
# attribute(s):
#       Min. 1st Qu. Median Mean 3rd Qu. Max.  NA's
# NO2  -8.1    3.02   5.66 8.39    10.4  197 16134
# dimension(s):
#          from    to       offset  delta              refsys
# time        1 8760 2017-01-01 UTC 1 hours            POSIXct
# station     1   74           NA      NA WGS 84 / UTM z...
#          point                              values
# time        NA                                NULL
# station   TRUE POINT (439814 ...,...,POINT (456668 ...
```

From this, we can compute the spatiotemporal variogram using

```
library(gstat)
```

```
v.st <- variogramST(NO2~1, no2.st[,1:(24*31)], tlags = 0:48,
    cores = getOption("mc.cores", 2))
```

which is shown in Figure 13.1.

To this sample variogram, we can fit a variogram model. One relatively flexible model we try here is the product-sum model (Gräler, Pebesma, and Heuvelink 2016), fitted by

```
# product-sum
prodSumModel <- vgmST("productSum",
    space = vgm(150, "Exp", 200000, 0),
    time = vgm(20, "Sph", 6, 0),
    k = 2)
#v.st$dist = v.st$dist / 1000
StAni <- estiStAni(v.st, c(0,200000))
```

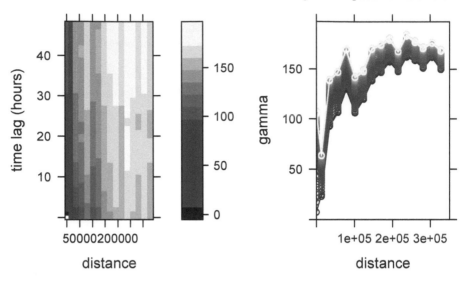

Figure 13.1: Spatiotemporal sample variogram for hourly NO_2 concentrations at rural background stations in Germany over 2027; in the right-hand side plot colour corresponds to time lag (yellow is later); distance in m

```
(fitProdSumModel <- fit.StVariogram(v.st, prodSumModel,
    fit.method = 7, stAni = StAni, method = "L-BFGS-B",
    control = list(parscale = c(1,100000,1,1,0.1,1,10)),
    lower = rep(0.0001, 7)))
# space component:
#    model    psill range
# 1   Nug    0.0166       0
# 2   Exp  152.7046   83590
# time component:
#    model    psill range
# 1   Nug    0.0001    0.00
# 2   Sph   25.5736    5.77
# k: 0.00397635996859073
```

and shown in Figure 13.2, which can also be plotted as wire frames, shown in Figure 13.3. Fitting this model is rather sensitive to the chosen parameters, which may be caused by the relatively small number (74) of monitoring network stations available.

Hints about the fitting strategy and alternative models for spatiotemporal variograms are given in Gräler, Pebesma, and Heuvelink (2016).

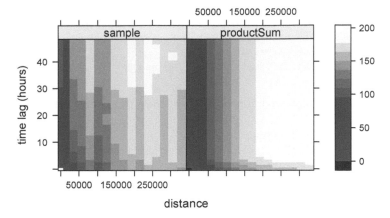

Figure 13.2: Product-sum model, fitted to the spatiotemporal sample variogram

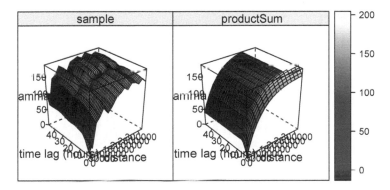

Figure 13.3: Wireframe plot of the fitted spatiotemporal variogram model

With this fitted model, and given the observations, we can carry out kriging or simulation at arbitrary points in space and time. For instance, we could estimate (or simulate) values in the time series that are now missing: this occurs regularly, and in Section 12.4 we used means over time series based on simply ignoring up to 25% of the observations: substituting these with estimated or simulated values based on neighbouring (in space and time) observations before computing yearly mean values seems a more reasonable approach.

More in general, we can estimate at arbitrary locations and time points, and we will illustrate this with predicting time series at particular locations and and predicting spatial slices (Gräler, Pebesma, and Heuvelink 2016). We can create a **stars** object for two randomly selected spatial points and all time instances by

```
set.seed(1331)
pt <- st_sample(de, 2)
t <- st_get_dimension_values(no2.st, 1)
```

```
st_as_stars(list(pts = matrix(1, length(t), length(pt)))) |>
    st_set_dimensions(names = c("time", "station")) |>
    st_set_dimensions("time", t) |>
    st_set_dimensions("station", pt) -> new_pt
```

and we obtain the spatiotemporal predictions at these two points using **krigeST** by

```
no2.st <- st_transform(no2.st, crs)
new_ts <- krigeST(NO2~1, data = no2.st["NO2"], newdata = new_pt,
        nmax = 50, stAni = StAni, modelList = fitProdSumModel,
        progress = FALSE)
```

where the results are shown in Figure 13.4.

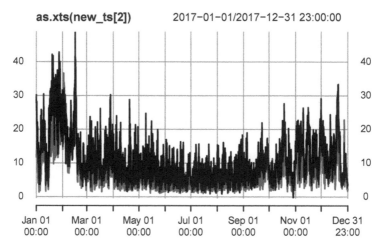

Figure 13.4: Time series plot of spatiotemporal predictions for two points

Alternatively, we can create spatiotemporal predictions for a set of time-stamped raster maps, evenly spaced over the year 2017, created by

```
st_bbox(de) |>
  st_as_stars(dx = 10000) |>
  st_crop(de) -> grd
d <- dim(grd)
t4 <- t[(1:4 - 0.5) * (3*24*30)]
st_as_stars(pts = array(1, c(d[1], d[2], time = length(t4)))) |>
    st_set_dimensions("time", t4) |>
    st_set_dimensions("x", st_get_dimension_values(grd, "x")) |>
    st_set_dimensions("y", st_get_dimension_values(grd, "y")) |>
    st_set_crs(crs) -> grd.st
```

and the subsequent predictions are obtained by

```
new_int <- krigeST(NO2~1, data = no2.st["NO2"], newdata = grd.st,
        nmax = 200, stAni = StAni, modelList = fitProdSumModel,
        progress = FALSE)
names(new_int)[2] = "NO2"
```

and shown in Figure 13.5.

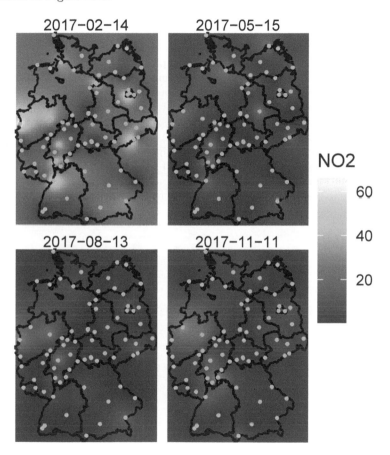

Figure 13.5: Spatiotemporal predictions for four selected time slices

A larger value for **nmax** was needed here to decrease the visible disturbance (sharp edges) caused by discrete neighbourhood selections, which are now done in space *and* time.

13.3.2   Irregular space time data

For the case where observations are collected at locations that vary constantly, or at fixed locations but without a common time basis, **stars** objects (vector data cubes) do not represent them well. Such irregular space time observations can be represented by **sftime** objects, provided by package **sftime** (Teickner, Pebesma, and Graeler 2022), which are essentially **sf** objects with a specified time column. An example of its uses is found in **demo(sftime)**, provided in package **gstat**.

13.4   Exercises

1. Which fraction of the stations is removed in Section 13.1 when the criterion applied that a station must be 75% complete?
2. From the hourly time series in **no2.st**, compute daily mean concentrations using **aggregate**, and compute the spatiotemporal variogram of this. How does it compare to the variogram of hourly values?
3. Carry out a spatiotemporal interpolation for daily mean values for the days corresponding to those shown in Figure 13.5, and compare the results.
4. Following the example in the demo scripts pointed at in Section 13.2, carry out a cokriging on the daily mean station data for the four days shown in Figure 13.5.
5. What are the differences of this latter approach to spatiotemporal kriging?

14

Proximity and Areal Data

Areal units of observation are very often used when simultaneous observations are aggregated within non-overlapping boundaries. The boundaries may be those of administrative entities and may be related to underlying spatial processes, such as commuting flows, but are usually arbitrary. If they do not match the underlying and unobserved spatial processes in one or more variables of interest, proximate areal units will contain parts of the underlying processes, engendering spatial autocorrelation. By proximity, we mean *closeness* in ways that make sense for the data generation processes thought to be involved. In cross-sectional geostatistical analysis with point support, measured distance makes sense for typical data generation processes. In similar analysis of areal data, sharing a border may make more sense, because that is what we do know, but we cannot measure the distance between the areas in as adequate a way.

By support of data we mean the physical size (length, area, volume) associated with an individual observational unit (measurement; see Chapter 5). It is possible to represent the support of areal data by a point, despite the fact that the data have polygonal support. The centroid of the polygon may be taken as a representative point, or the centroid of the largest polygon in a multi-polygon object. When data with intrinsic point support are treated as areal data, the change of support goes the other way, from the known point to a non-overlapping tessellation such as a Voronoi diagram or Dirichlet tessellation or Thiessen polygons often through a Delaunay triangulation using projected coordinates. Here, different metrics may also be chosen, or distances measured on a network rather than on the plane. There is also a literature using weighted Voronoi diagrams in local spatial analysis (see for example Boots and Okabe 2007; Okabe et al. 2008; She et al. 2015).

When the intrinsic support of the data is represented as points, but the underlying process is between proximate observations rather than driven chiefly by distance between observations, the data may be aggregate counts or totals (polling stations, retail turnover) or represent a directly observed characteristic of the observation (opening hours of the polling station). Obviously, the risk of misrepresenting the footprint of the underlying spatial processes remains in all of these cases, not least because the observations are taken as encompassing the entirety of the underlying process in the case of tessellation of the whole area of interest. This is distinct from the geostatistical setting in which observations are rather samples taken using some scheme within the area of interest. It is also partly distinct from the practice of taking areal sample plots within the area of interest but covering only a small proportion of the area, typically used in ecological and environmental research.

In order to explore and analyse areal data of these kinds in Chapters -Chapter 15 -
-Chapter 17, methods are needed to represent the proximity of observations. This
chapter considers a subset of such methods, where the spatial processes are considered
as working through proximity understood in the first instance as contiguity, as a
graph linking observations taken as neighbours. This graph is typically undirected
and unweighted, but may be directed and/or weighted in certain settings, which then
leads to further issues with regard to symmetry. In principle, proximity would be
expected to operate symmetrically in space, that is that the influence of i on j and
of j on i based on their relative positions should be equivalent. Edge effects are not
considered in standard treatments.

14.1   Representing proximity in spdep

Handling spatial autocorrelation using relationships to neighbours on a graph takes
the graph as given, chosen by the analyst. This differs from the geostatistical approach
in which the analyst chooses the binning of the empirical variogram and function
used, and then the way the variogram is fitted. Both involve a priori choices, but
represent the underlying correlation in different ways (Wall 2004). In Bavaud (1998)
and work citing his contribution, attempts have been made to place graph-based
neighbours in a broader context.

One issue arising in the creation of objects representing neighbourhood relationships
is that of no-neighbour areal units (Bivand and Portnov 2004). Islands or units
separated by rivers may not be recognised as neighbours when the units have areal
support and when using topological relationships such as shared boundaries. In some
settings, for example **mrf** (Markov Random Field) terms in **mgcv::gam** and similar
model fitting functions, undirected connected graphs are required, which is violated
when there are disconnected subgraphs.

No-neighbour observations can also occur when a distance threshold is used between
points, where the threshold is smaller than the maximum nearest neighbour distance.
Shared boundary contiguities are not affected by using geographical, unprojected
coordinates, but all point-based approaches use distance in one way or another, and
need to calculate distances in an appropriate way.

The **spdep** package provides an **nb** class for neighbours, a list of length equal to the
number of observations, with integer vector components. No-neighbours are encoded
as an integer vector with a single element **0L**, and observations with neighbours
as sorted integer vectors containing values in **1L:n** pointing to the neighbouring
observations. This is a typical row-oriented sparse representation of neighbours.
spdep provides many ways of constructing **nb** objects, and the representation and
construction functions are widely used in other packages.

spdep builds on the **nb** representation (undirected or directed graphs) with the
listw object, a list with three components, an **nb** object, a matching list of numerical
weights, and a single element character vector containing the single letter name of
the way in which the weights were calculated. The most frequently used approach

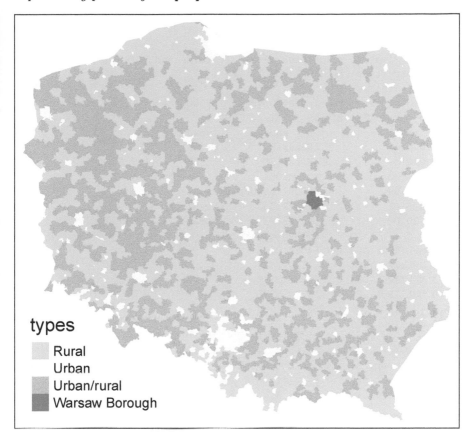

types

Rural
Urban
Urban/rural
Warsaw Borough

Figure 14.1: Polish municipality types 2015

in the social sciences is calculating weights by row standardisation, so that all the non-zero weights for one observation will be the inverse of the cardinality of its set of neighbours (`1/card(nb)[i]`).

We will be using election data from the 2015 Polish presidential election in this chapter, with 2495 municipalities and Warsaw boroughs (see Figure 14.1) for a **tmap** map (Section 8.5) of the municipality types, and complete count data from polling stations aggregated to these areal units. The data are an **sf sf** object:

```
library(sf)
# Linking to GEOS 3.11.1, GDAL 3.6.2, PROJ 9.1.1; sf_use_s2()
# is TRUE

data(pol_pres15, package = "spDataLarge")
pol_pres15 |>
```

```
        subset(select = c(TERYT, name, types)) |>
        head()
# Simple feature collection with 6 features and 3 fields
# Geometry type: MULTIPOLYGON
# Dimension:     XY
# Bounding box:  xmin: 235000 ymin: 367000 xmax: 281000 ymax: 413000
# Projected CRS: ETRS89 / Poland CS92
#    TERYT              name          types
# 1 020101         BOLESŁAWIEC        Urban
# 2 020102         BOLESŁAWIEC        Rural
# 3 020103            GROMADKA        Rural
# 4 020104         NOWOGRODZIEC  Urban/rural
# 5 020105           OSIECZNICA        Rural
# 6 020106 WARTA BOLESŁAWIECKA        Rural
#                        geometry
# 1 MULTIPOLYGON (((261089 3855...
# 2 MULTIPOLYGON (((254150 3837...
# 3 MULTIPOLYGON (((275346 3846...
# 4 MULTIPOLYGON (((251770 3770...
# 5 MULTIPOLYGON (((263424 4060...
# 6 MULTIPOLYGON (((267031 3870...


library(tmap, warn.conflicts = FALSE)
tm_shape(pol_pres15) + tm_fill("types")
```

For safety's sake, we impose topological validity:

```
if (!all(st_is_valid(pol_pres15)))
        pol_pres15 <- st_make_valid(pol_pres15)
```

Between early 2002 and April 2019, **spdep** contained functions for constructing and handling neighbour and spatial weights objects, tests for spatial autocorrelation, and model fitting functions. The latter have been split out into **spatialreg**, and will be discussed in subsequent chapters. **spdep** (Bivand 2022c) now accommodates objects represented using **sf** classes and **sp** classes directly.

```
library(spdep) |> suppressPackageStartupMessages()
```

14.2 Contiguous neighbours

The `poly2nb` function in **spdep** takes the boundary points making up the polygon boundaries in the object passed as the `pl=` argument, typically an `"sf"` or `"sfc"` object with `"POLYGON"` or `"MULTIPOLYGON"` geometries. For each observation, the function checks whether at least one (**queen=TRUE**, default), or at least two (rook, **queen=FALSE**) points are within **snap=** distance units of each other. The distances are planar in the raw coordinate units, ignoring geographical projections. Once the required number of sufficiently close points is found, the search is stopped.

```
args(poly2nb)
```

```
#  function (pl, row.names = NULL, snap =
#     sqrt(.Machine$double.eps), queen = TRUE, useC = TRUE,
#     foundInBox = NULL)
```

From **spdep** 1.1-7, the **sf** package GEOS interface is used within **poly2nb** to find the candidate neighbours and populate **foundInBox** internally. In this case, the use of spatial indexing (STRtree queries) in GEOS through **sf** is the default:

```
pol_pres15 |> poly2nb(queen = TRUE) -> nb_q
```

The print method shows the summary structure of the neighbour object:

```
nb_q
# Neighbour list object:
# Number of regions: 2495
# Number of nonzero links: 14242
# Percentage nonzero weights: 0.229
# Average number of links: 5.71
```

From **sf** version 1.0-0, the **s2** package (Dunnington, Pebesma, and Rubak 2023) is used by default for spherical geometries, as **st_intersects** used in **poly2nb** passes calculation to **s2::s2_intersects_matrix** (see Chapter 4). From **spdep** version 1.1-9, if **sf_use_s2()** is TRUE, spherical intersection is used to find candidate neighbours; as with GEOS, the underlying **s2** library uses fast spatial indexing.

```
old_use_s2 <- sf_use_s2()
```

```
sf_use_s2(TRUE)
```

```
(pol_pres15 |> st_transform("OGC:CRS84") -> pol_pres15_ll) |>
    poly2nb(queen = TRUE) -> nb_q_s2
```

Spherical and planar intersection of the input polygons yield the same contiguity neighbours in this case; in both cases valid input geometries are desirable:

```
all.equal(nb_q, nb_q_s2, check.attributes=FALSE)
# [1] TRUE
```

Note that **nb** objects record both symmetric neighbour relationships i to j and j to i, because these objects admit asymmetric relationships as well, but these duplications are not needed for object construction.

Most of the **spdep** functions for constructing neighbour objects take a **row.names=** argument, the value of which is stored as a **region.id** attribute. If not given, the values are taken from **row.names()** of the first argument. These can be used to check that the neighbours object is in the same order as data. If **nb** objects are subsetted, the indices change to continue to be within **1:length(subsetted_nb)**, but the **region.id** attribute values point back to the object from which it was constructed. This is used in out-of-sample prediction from spatial regression models discussed briefly in Section 17.4.

We can also check that this undirected graph is connected using the **n.comp.nb** function; while some model estimation techniques do not support graphs that are not connected, it is helpful to be aware of possible problems (Freni-Sterrantino, Ventrucci, and Rue 2018):

```
(nb_q |> n.comp.nb())$nc
# [1] 1
```

This approach is equivalent to treating the neighbour object as a graph and using graph analysis on that graph (Csardi and Nepusz 2006; Nepusz 2022), by first coercing to a binary sparse matrix (Bates, Maechler, and Jagan 2022):

```
library(Matrix, warn.conflicts = FALSE)
library(spatialreg, warn.conflicts = FALSE)
nb_q |>
    nb2listw(style = "B") |>
    as("CsparseMatrix") -> smat
library(igraph, warn.conflicts = FALSE)
(smat |> graph.adjacency() -> g1) |>
    count_components()
# [1] 1
```

Neighbour objects may be exported and imported in GAL format for exchange with other software, using **write.nb.gal** and **read.gal**:

```
tf <- tempfile(fileext = ".gal")
write.nb.gal(nb_q, tf)
```

14.3 Graph-based neighbours

If areal units are an appropriate representation, but only points on the plane have been observed, contiguity relationships may be approximated using graph-based neighbours. In this case, the imputed boundaries tessellate the plane such that points closer to one observation than any other fall within its polygon. The simplest form is by using triangulation, here using the **deldir** function in the **deldir** package. Because the function returns from i and to j identifiers, it is easy to construct a long representation of a **listw** object, as used in the S-Plus SpatialStats module and the **sn2listw** function internally to construct an **nb** object (ragged wide representation). Alternatives such as GEOS often fail to return sufficient information to permit the neighbours to be identified.

The output of these functions is then converted to the **nb** representation using **graph2nb**, with the possible use of the **sym=** argument to coerce to symmetry. We take the centroids of the largest component polygon for each observation as the point representation; population-weighted centroids might have been a better choice if they were available:

```
pol_pres15 |>
    st_geometry() |>
    st_centroid(of_largest_polygon = TRUE) -> coords
(coords |> tri2nb() -> nb_tri)
# Neighbour list object:
# Number of regions: 2495
# Number of nonzero links: 14930
# Percentage nonzero weights: 0.24
# Average number of links: 5.98
```

The average number of neighbours is similar to the Queen boundary contiguity case, but if we look at the distribution of edge lengths using **nbdists()**, we can see that although the upper quartile is about 15 km, the maximum is almost 300 km, an edge along much of one side of the convex hull. The short minimum distance is also of interest, as many centroids of urban municipalities are very close to the centroids of their surrounding rural counterparts.

```
nb_tri |>
    nbdists(coords) |>
    unlist() |>
    summary()
#    Min. 1st Qu.  Median    Mean 3rd Qu.    Max.
#     247    9847   12151   13485   14994  296974
```

Triangulated neighbours also yield a connected graph:

```
(nb_tri |> n.comp.nb())$nc
# [1] 1
```

Graph-based approaches include **soi.graph** - discussed here, **relativeneigh** and **gabrielneigh**.

The Sphere of Influence **soi.graph** function takes triangulated neighbours and prunes off neighbour relationships represented by edges that are unusually long for each point, especially around the convex hull (Avis and Horton 1985).

```
(nb_tri |>
        soi.graph(coords) |>
        graph2nb() -> nb_soi)
# Neighbour list object:
# Number of regions: 2495
# Number of nonzero links: 12792
# Percentage nonzero weights: 0.205
# Average number of links: 5.13
```

Unpicking the triangulated neighbours does however remove the connected character of the underlying graph:

```
(nb_soi |> n.comp.nb() -> n_comp)$nc
# [1] 16
```

The algorithm has stripped out longer edges leading to urban and rural municipality pairs where their centroids are very close to each other because the rural ones completely surround the urban, giving 15 pairs of neighbours unconnected to the main graph:

```
table(n_comp$comp.id)
#
#    1    2    3    4    5    6    7    8    9   10   11   12
# 2465    2    2    2    2    2    2    2    2    2    2    2
#   13   14   15   16
#    2    2    2    2
```

The largest length edges along the convex hull have been removed, but "holes" have appeared where the unconnected pairs of neighbours have appeared. The differences between **nb_tri** and **nb_soi** are shown in orange in Figure 14.2.

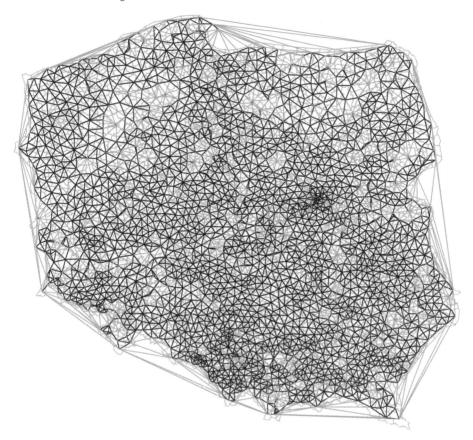

Figure 14.2: Triangulated (orange + black) and sphere of influence neighbours (black); apparent holes appear for sphere of influence neighbours where an urban municipality is surrounded by a dominant rural municipality (see Figure 14.1)

14.4 Distance-based neighbours

Distance-based neighbours can be constructed using **dnearneigh**, with a distance band with lower **d1=** and upper **d2=** bounds controlled by the **bounds=** argument. If spherical coordinates are used and either specified in the coordinates object x or with x as a two-column matrix and **longlat=TRUE**, great circle distances in kilometre will be calculated assuming the WGS84 reference ellipsoid, or if **use_s2=TRUE** (the default value) using the spheroid (see Chapter 4). If **dwithin=** is **FALSE** and the version of **s2** is greater than **1.0.7**, **s2_closest_edges** may be used, if **TRUE** and **use_s2=TRUE**, **s2_dwithin_matrix** is used; both of these methods use fast spherical spatial indexing, but because **s2_closest_edges** takes minimum and maximum bounds, it only needs one pass in the R code of **dnearneigh**.

Arguments have been added to use functionality in the **dbscan** package (Hahsler and Piekenbrock 2022) for finding neighbours using planar spatial indexing in two or three dimensions by default, and not to test the symmetry of the output neighbour object. In addition, three arguments relate to the use of spherical geometry distance measurements.

The **knearneigh** function for k-nearest neighbours returns a **knn** object, converted to an **nb** object using **knn2nb**. It can also use great circle distances, not least because nearest neighbours may differ when unprojected coordinates are treated as planar. **k=** should be a small number. For projected coordinates, the **dbscan** package is used to compute nearest neighbours more efficiently. Note that **nb** objects constructed in this way are most unlikely to be symmetric hence **knn2nb** has a **sym=** argument to permit the imposition of symmetry, which will mean that all units have at least **k=** neighbours, not that all units will have exactly **k=** neighbours. When **sf_use_s2()** is TRUE, **knearneigh** will use fast spherical spatial indexing when the input object is of class "**sf**" or "**sfc**".

The **nbdists** function returns the length of neighbour relationship edges in the units of the coordinates if the coordinates are projected, in kilometre otherwise. In order to set the upper limit for distance bands, one may first find the maximum first nearest neighbour distance, using **unlist** to remove the list structure of the returned object. When **sf_use_s2()** is TRUE, **nbdists** will use fast spherical distance calculations when the input object is of class "**sf**" or "**sfc**".

```
coords |>
    knearneigh(k = 1) |>
    knn2nb() |>
    nbdists(coords) |>
    unlist() |>
    summary()
#    Min. 1st Qu.  Median   Mean 3rd Qu.    Max.
#     247    6663    8538    8275   10124   17979
```

Here the largest first nearest neighbour distance is just under 18 km, so using this as the upper threshold gives certainty that all units will have at least one neighbour:

```
coords |> dnearneigh(0, 18000) -> nb_d18
```

For this moderate number of observations, use of spatial indexing does not yield advantages in run times:

```
coords |> dnearneigh(0, 18000, use_kd_tree = FALSE) -> nb_d18a
```

and the output objects are the same:

```
all.equal(nb_d18, nb_d18a, check.attributes = FALSE)
# [1] TRUE
```

```
nb_d18
# Neighbour list object:
# Number of regions: 2495
# Number of nonzero links: 20358
# Percentage nonzero weights: 0.327
# Average number of links: 8.16
```

However, even though there are no no-neighbour observations (their presence is reported by the print method for **nb** objects), the graph is not connected, as a pair of observations are each others' only neighbours.

```
(nb_d18 |> n.comp.nb() -> n_comp)$nc
# [1] 2
```

```
table(n_comp$comp.id)
#
#    1    2
# 2493    2
```

Adding 300 m to the threshold gives us a neighbour object with no no-neighbour units, and all units can be reached from all others across the graph.

```
(coords |> dnearneigh(0, 18300) -> nb_d183)
# Neighbour list object:
# Number of regions: 2495
# Number of nonzero links: 21086
# Percentage nonzero weights: 0.339
# Average number of links: 8.45
```

```
(nb_d183 |> n.comp.nb())$nc
# [1] 1
```

One characteristic of distance-based neighbours is that more densely settled areas, with units which are smaller in terms of area, have higher neighbour counts (Warsaw boroughs are much smaller on average, but have almost 30 neighbours for this distance criterion). Having many neighbours smooths the neighbour relationship across more neighbours.

For use later, we also construct a neighbour object with no-neighbour units, using a threshold of 16 km:

```
(coords |> dnearneigh(0, 16000) -> nb_d16)
# Neighbour list object:
# Number of regions: 2495
# Number of nonzero links: 15850
```

```
# Percentage nonzero weights: 0.255
# Average number of links: 6.35
# 7 regions with no links:
# 569 1371 1522 2374 2385 2473 2474
```

It is possible to control the numbers of neighbours directly using *k*-nearest neighbours, either accepting asymmetric neighbours:

```
((coords |> knearneigh(k = 6) -> knn_k6) |> knn2nb() -> nb_k6)
# Neighbour list object:
# Number of regions: 2495
# Number of nonzero links: 14970
# Percentage nonzero weights: 0.24
# Average number of links: 6
# Non-symmetric neighbours list
```

or imposing symmetry:

```
(knn_k6 |> knn2nb(sym = TRUE) -> nb_k6s)
# Neighbour list object:
# Number of regions: 2495
# Number of nonzero links: 16810
# Percentage nonzero weights: 0.27
# Average number of links: 6.74
```

Here the size of `k=` is sufficient to ensure connectedness, although the graph is not planar as edges cross at locations other than nodes, which is not the case for contiguous or graph-based neighbours.

```
(nb_k6s |> n.comp.nb())$nc
# [1] 1
```

In the case of points on the sphere (see Chapter 4), the output of **st_centroid** will differ, so rather than inverse projecting the points, we extract points as geographical coordinates from the inverse projected polygon geometries:

```
old_use_s2 <- sf_use_s2()
```

```
sf_use_s2(TRUE)
```

```
pol_pres15_ll |>
    st_geometry() |>
    st_centroid(of_largest_polygon = TRUE) -> coords_ll
```

For spherical coordinates, distance bounds are in kilometres:

```
(coords_ll |> dnearneigh(0, 18.3, use_s2 = TRUE,
                        dwithin = TRUE) -> nb_d183_ll)
# Neighbour list object:
# Number of regions: 2495
# Number of nonzero links: 21140
# Percentage nonzero weights: 0.34
# Average number of links: 8.47
```

These neighbours differ from the spherical 18.3 km neighbours as would be expected:

```
isTRUE(all.equal(nb_d183, nb_d183_ll, check.attributes = FALSE))
# [1] FALSE
```

If **s2** providing faster distance neighbour indexing is available, by default `s2_closest_edges` will be used for geographical coordinates:

```
(coords_ll |> dnearneigh(0, 18.3) -> nb_d183_llce)
# Neighbour list object:
# Number of regions: 2495
# Number of nonzero links: 21140
# Percentage nonzero weights: 0.34
# Average number of links: 8.47
```

where the two **s2**-based neighbour objects are the same:

```
isTRUE(all.equal(nb_d183_llce, nb_d183_ll,
                 check.attributes = FALSE))
# [1] TRUE
```

Fast spherical spatial indexing in **s2** is used to find k nearest neighbours:

```
(coords_ll |> knearneigh(k = 6) |> knn2nb() -> nb_k6_ll)
# Neighbour list object:
# Number of regions: 2495
# Number of nonzero links: 14970
# Percentage nonzero weights: 0.24
# Average number of links: 6
# Non-symmetric neighbours list
```

These neighbours differ from the planar k=6 nearest neighbours as would be expected, but will also differ slightly from legacy brute-force ellipsoid distances:

```
isTRUE(all.equal(nb_k6, nb_k6_ll, check.attributes = FALSE))
# [1] FALSE
```

The **nbdists** function also uses **s2** to find distances on the sphere when the `"sf"` or `"sfc"`input object is in geographical coordinates (distances returned in kilometres):

```
nb_q |> nbdists(coords_ll) |> unlist() |> summary()
#    Min. 1st Qu.  Median   Mean 3rd Qu.    Max.
#     0.2     9.8    12.2   12.6    15.1    33.0
```

These differ a little for the same weights object when planar coordinates are used (distances returned in the metric of the points for planar geometries and kilometres for ellipsoidal and spherical geometries):

```
nb_q |> nbdists(coords) |> unlist() |> summary()
#    Min. 1st Qu.  Median   Mean 3rd Qu.    Max.
#     247    9822   12173  12651   15117   33102
```

```
sf_use_s2(old_use_s2)
```

14.5   Weights specification

Once neighbour objects are available, further choices need to be made in specifying the weights objects. The **nb2listw** function is used to create a **listw** weights object with an **nb** object, a matching list of weights vectors, and a style specification. Because handling no-neighbour observations now begins to matter, the **zero.policy=** argument is introduced. By default, this is **FALSE**, indicating that no-neighbour observations will cause an error, as the spatially lagged value for an observation with no neighbours is not available. By convention, zero is substituted for the lagged value, as the cross-product of a vector of zero-valued weights and a data vector, hence the name of **zero.policy**.

```
args(nb2listw)
```

```
#  function (neighbours, glist = NULL, style = "W", zero.policy
#     = NULL)
```

We will be using the helper function **spweights.constants** below to show some consequences of varying style choices. It returns constants for a **listw** object, n is the number of observations, **n1** to **n3** are $n - 1, \ldots$, **nn** is n^2 and S_0, S_1 and S_2 are constants, S_0 being the sum of the weights. There is a full discussion of the constants in Bivand and Wong (2018).

```
args(spweights.constants)
```

```
#  function (listw, zero.policy = NULL, adjust.n = TRUE)
```

The "B" binary style gives a weight of unity to each neighbour relationship, and typically up-weights units with no boundaries on the edge of the study area, having a higher count of neighbours.

```
(nb_q |>
    nb2listw(style = "B") -> lw_q_B) |>
    spweights.constants() |>
    data.frame() |>
    subset(select = c(n, S0, S1, S2))
#      n    S0    S1    S2
# 1 2495 14242 28484 357280
```

The "W" row-standardised style up-weights units around the edge of the study area that necessarily have fewer neighbours. This style first gives a weight of unity to each neighbour relationship, then it divides these weights by the per unit sums of weights. Naturally this leads to division by zero where there are no neighbours, a not-a-number result, unless the chosen policy is to permit no-neighbour observations. We can see that S_0 is now equal to n.

```
(nb_q |>
        nb2listw(style = "W") -> lw_q_W) |>
    spweights.constants() |>
    data.frame() |>
    subset(select = c(n, S0, S1, S2))
#      n   S0  S1    S2
# 1 2495 2495 958 10406
```

Inverse distance weights are used in a number of scientific fields. Some use dense inverse distance matrices, but many of the inverse distances are close to zero, have little practical contribution, especially as the spatial process matrix is itself dense. Inverse distance weights may be constructed by taking the lengths of edges, changing units to avoid most weights being too large or small (here from metre to kilometre), taking the inverse, and passing through the **glist=** argument to **nb2listw**:

```
nb_d183 |>
    nbdists(coords) |>
    lapply(function(x) 1/(x/1000)) -> gwts
(nb_d183 |> nb2listw(glist=gwts, style="B") -> lw_d183_idw_B) |>
    spweights.constants() |>
    data.frame() |>
    subset(select=c(n, S0, S1, S2))
#      n   S0  S1   S2
# 1 2495 1841 534 7265
```

No-neighbour handling is by default to prevent the construction of a weights object, making the analyst take a position on how to proceed.

```
try(nb_d16 |> nb2listw(style="B") -> lw_d16_B)
# Error in nb2listw(nb_d16, style = "B") : Empty neighbour sets found
```

Use can be made of the **zero.policy=** argument to many functions used with **nb** and **listw** objects.

```
nb_d16 |>
    nb2listw(style="B", zero.policy=TRUE) |>
    spweights.constants(zero.policy=TRUE) |>
    data.frame() |>
    subset(select=c(n, S0, S1, S2))
#      n    S0    S1     S2
# 1 2488 15850 31700 506480
```

Note that by default the **adjust.n=** argument to **spweights.constants** is set by default to **TRUE**, subtracting the count of no-neighbour observations from the observation count, so n is smaller with possible consequences for inference. The complete count can be retrieved by changing the argument.

14.6   Higher order neighbours

We recall the characteristics of the neighbour object based on Queen contiguities:

```
nb_q
# Neighbour list object:
# Number of regions: 2495
# Number of nonzero links: 14242
# Percentage nonzero weights: 0.229
# Average number of links: 5.71
```

If we wish to create an object showing i to k neighbours, where i is a neighbour of j, and j in turn is a neighbour of k, so taking two steps on the neighbour graph, we can use **nblag**, which automatically removes i to i self-neighbours:

```
(nb_q |> nblag(2) -> nb_q2)[[2]]
# Neighbour list object:
# Number of regions: 2495
# Number of nonzero links: 32930
# Percentage nonzero weights: 0.529
# Average number of links: 13.2
```

The **nblag_cumul** function cumulates the list of neighbours for the whole list of lags:

```
nblag_cumul(nb_q2)
# Neighbour list object:
# Number of regions: 2495
# Number of nonzero links: 47172
# Percentage nonzero weights: 0.758
# Average number of links: 18.9
```

while the set operation **union.nb** takes two objects, giving here the same outcome:

```
union.nb(nb_q2[[2]], nb_q2[[1]])
# Neighbour list object:
# Number of regions: 2495
# Number of nonzero links: 47172
# Percentage nonzero weights: 0.758
# Average number of links: 18.9
```

Returning to the graph representation of the same neighbour object, we can ask how many steps might be needed to traverse the graph:

```
diameter(g1)
# [1] 52
```

We step out from each observation across the graph to establish the number of steps needed to reach each other observation by the shortest path (creating an $n \times n$ matrix **sps**), once again finding the same maximum count.

```
g1 |> shortest.paths() -> sps
(sps |> apply(2, max) -> spmax) |> max()
# [1] 52
```

The municipality with the maximum count is called Lutowiska, close to the Ukrainian border in the far south east of the country:

```
mr <- which.max(spmax)
pol_pres15$name0[mr]
# [1] "Lutowiska"
```

Figure 14.3 shows that contiguity neighbours represent the same kinds of relationships with other observations as distance. Some approaches prefer distance neighbours on the basis that, for example, inverse distance neighbours show clearly how all observations are related to each other. However, the development of tests for spatial autocorrelation and spatial regression models has involved the inverse of a spatial process model, which in turn can be represented as the sum of a power series of the product of a coefficient and a spatial weights matrix, intrinsically acknowledging the relationships of all observations with all other observations. Sparse contiguity neighbour objects accommodate rich dependency structures without the need to make the structures explicit.

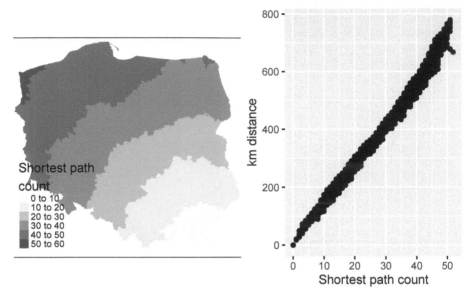

Figure 14.3: Relationship of shortest paths to distance for Lutowiska; left panel: shortest path counts from Lutowiska; right panel: plot of shortest paths from Lutowiska to other observations, and distances from Lutowiska to other observations

14.7 Exercises

1. Which kinds of geometry support are appropriate for which functions creating neighbour objects?
2. Which functions creating neighbour objects are only appropriate for planar representations?
3. What difference might the choice of **rook** rather than **queen** contiguities make on a chessboard?
4. What are the relationships between neighbour set cardinalities (neighbour counts) and row-standardised weights, and how do they open analyses up to edge effects? Use the chessboard you constructed in exercise 3 for both **rook** and **queen** neighbours.

15

Measures of Spatial Autocorrelation

When analysing areal data, it has long been recognised that, if present, spatial autocorrelation changes how we may infer, relative to the default assumption of independent observations. In the presence of spatial autocorrelation, we can predict the values of observation i from the values observed at $j \in N_i$, the set of its proximate neighbours. Early results (Moran 1948; Geary 1954) entered into research practice gradually, for example the social sciences (Duncan, Cuzzort, and Duncan 1961). These results were then collated and extended to yield a set of basic tools of analysis (Cliff and Ord 1973, 1981).

Cliff and Ord (1973) generalised and extended the expression of the spatial weights matrix representation as part of the framework for establishing the distribution theory for join-count, Moran's I and Geary's C statistics. This development of what have become known as global measures, returning a single value of autocorrelation for the total study area, has been supplemented by local measures returning values for each areal unit (Getis and Ord 1992; Anselin 1995).

The measures offered by the **spdep** package have been written partly to provide implementations, but also to permit the comparative investigation of these measures and their implementation. For this reason, the implementations are written in R rather than compiled code, and they are generally slower but more flexible than implementations in the newly released **rgeoda** package (Li and Anselin 2021; Anselin, Li, and Koschinsky 2021).

15.1 Measures and process misspecification

It is not and has never been the case that Tobler's first law of geography, "Everything is related to everything else, but near things are more related than distant things", always holds absolutely. This is and has always been an oversimplification, disguising possible underlying entitation, support, and other misspecification problems. Are the units of observation appropriate for the scale of the underlying spatial process? Could the spatial patterning of the variable of interest for the chosen entitation be accounted for by another variable?

Tobler (1970) was published in the same special issue of *Economic Geography* as Olsson (1970), but Olsson does grasp the important point that spatial autocorrelation is not inherent in spatial phenomena, but often, is engendered by inappropriate entitation, by omitted variables and/or inappropriate functional form. The key quote from Olsson is on p. 228:

> The existence of such autocorrelations makes it tempting to agree with Tobler (1970, 236 [the original refers to the pagination of a conference paper]) that 'everything is related to everything else, but near things are more related than distant things.' On the other hand, the fact that the autocorrelations seem to hide systematic specification errors suggests that the elevation of this statement to the status of 'the first law of geography' is at best premature. At worst, the statement may represent the spatial variant of the post hoc fallacy, which would mean that coincidence has been mistaken for a causal relation.

The status of the "first law" is very similar to the belief that John Snow induced from a map the cause of cholera as water-borne. It may be a good way of selling GIS, but it is inaccurate: Snow had a strong working hypothesis prior to visiting Soho, and the map was prepared after the Broad Street pump was disabled as documentation that his hypothesis held (Brody et al. 2000).

Measures of spatial autocorrelation unfortunately pick up other misspecifications in the way that we model data (Schabenberger and Gotway 2005; McMillen 2003). For reference, Moran's I is given as (Cliff and Ord 1981, 17):

$$I = \frac{n \sum_{(2)} w_{ij} z_i z_j}{S_0 \sum_{i=1}^{n} z_i^2}$$

where $x_i, i = 1, \ldots, n$ are n observations on the numeric variable of interest, $z_i = x_i - \bar{x}$, $\bar{x} = \sum_{i=1}^{n} x_i / n$, $\sum_{(2)} = \sum_{i=1}^{n} \sum_{j=1}^{n} {}_{i \neq j}$, w_{ij} are the spatial weights, and $S_0 = \sum_{(2)} w_{ij}$. First we test a random variable using the Moran test, here under the normality assumption (argument **randomisation=FALSE**, default **TRUE**). Inference is made on the statistic $Z(I) = \frac{I - E(I)}{\sqrt{\mathrm{Var}(I)}}$, the z-value compared with the Normal distribution for $E(I)$ and $\mathrm{Var}(I)$ for the chosen assumptions; this **x** does not show spatial autocorrelation with these spatial weights:

```
library(spdep) |> suppressPackageStartupMessages()
library(parallel)
glance_htest <- function(ht) c(ht$estimate,
    "Std deviate" = unname(ht$statistic),
    "p.value" = unname(ht$p.value))
set.seed(1)
(pol_pres15 |>
    nrow() |>
    rnorm() -> x) |>
    moran.test(lw_q_B, randomisation = FALSE,
               alternative = "two.sided") |>
    glance_htest()
# Moran I statistic      Expectation            Variance
#        -0.004772        -0.000401            0.000140
```

```
#       Std deviate             p.value
#        -0.369320             0.711889
```

The test however detects quite strong positive spatial autocorrelation when we insert a gentle trend into the data, but omit to include it in the mean model, thus creating a missing variable problem but finding spatial autocorrelation instead:

```
beta <- 0.0015
coords |>
    st_coordinates() |>
    subset(select = 1, drop = TRUE) |>
    (function(x) x/1000)() -> t
(x + beta * t -> x_t) |>
    moran.test(lw_q_B, randomisation = FALSE,
               alternative = "two.sided") |>
    glance_htest()
# Moran I statistic       Expectation            Variance
#          0.043403         -0.000401            0.000140
#       Std deviate             p.value
#          3.701491           0.000214
```

If we test the residuals of a linear model including the trend, the apparent spatial autocorrelation disappears:

```
lm(x_t ~ t) |>
    lm.morantest(lw_q_B, alternative = "two.sided") |>
    glance_htest()
# Observed Moran I        Expectation            Variance
#         -0.004777         -0.000789            0.000140
#       Std deviate             p.value
#         -0.337306           0.735886
```

A comparison of implementations of measures of spatial autocorrelation shows that a wide range of measures is available in R in a number of packages, chiefly in the **spdep** package (Bivand 2022c), and that differences from other implementations can be attributed to design decisions (Bivand and Wong 2018). The **spdep** package also includes the only implementations of exact and saddlepoint approximations to global and local Moran's I for regression residuals (Tiefelsdorf 2002; Bivand, Müller, and Reder 2009).

15.2 Global measures

Global measures consider the average level of spatial autocorrelation across all observations; they can of course be biased (as most spatial statistics) by edge effects where important spatial process components fall outside the study area.

15.2.1 Join-count tests for categorical data

We will begin by examining join-count statistics, where `joincount.test` takes a `"factor"` vector of values `fx=` and a `listw` object, and returns a list of `htest` (hypothesis test) objects defined in the **stats** package, one `htest` object for each level of the `fx=` argument. The observed counts are of neighbours with the same factor levels, known as same-colour joins.

```
args(joincount.test)
```

```
# function (fx, listw, zero.policy = NULL, alternative =
#     "greater", sampling = "nonfree", spChk = NULL, adjust.n =
#     TRUE)
```

The function takes an **alternative=** argument for hypothesis testing, a **sampling=** argument showing the basis for the construction of the variance of the measure, where the default `"nonfree"` choice corresponds to analytical permutation; the **spChk=** argument is retained for backward compatibility. For reference, the counts of factor levels for the type of municipality or Warsaw borough are:

```
(pol_pres15 |>
        st_drop_geometry() |>
        subset(select = types, drop = TRUE) -> Types) |>
    table()
#
#      Rural       Urban   Urban/rural Warsaw Borough
#       1563         303           611             18
```

Since there are four levels, we rearrange the list of **htest** objects to give a matrix of estimated results. The observed same-colour join-counts are tabulated with their expectations based on the counts of levels of the input factor, so that few joins would be expected between for example Warsaw boroughs, because there are very few of them. The variance calculation uses the underlying constants of the chosen `listw` object and the counts of levels of the input factor. The z-value is obtained in the usual way by dividing the difference between the observed and expected join-counts by the square root of the variance.

The join-count test was subsequently adapted for multi-colour join-counts (Upton and Fingleton 1985). The implementation as `joincount.multi` in **spdep** returns a table based on non-free sampling, and does not report p-values.

```
Types |> joincount.multi(listw = lw_q_B)
#                                 Joincount Expected Variance
# Rural:Rural                      3087.000 2793.920 1126.534
# Urban:Urban                       110.000  104.719   93.299
# Urban/rural:Urban/rural           656.000  426.526  331.759
# Warsaw Borough:Warsaw Borough      41.000    0.350    0.347
# Urban:Rural                       668.000 1083.941  708.209
```

```
# Urban/rural:Rural                    2359.000 2185.769 1267.131
# Urban/rural:Urban                     171.000  423.729  352.190
# Warsaw Borough:Rural                   12.000   64.393   46.460
# Warsaw Borough:Urban                    9.000   12.483   11.758
# Warsaw Borough:Urban/rural              8.000   25.172   22.354
# Jtot                                 3227.000 3795.486 1496.398
#                                      z-value
# Rural:Rural                             8.73
# Urban:Urban                             0.55
# Urban/rural:Urban/rural                12.60
# Warsaw Borough:Warsaw Borough          68.96
# Urban:Rural                           -15.63
# Urban/rural:Rural                       4.87
# Urban/rural:Urban                     -13.47
# Warsaw Borough:Rural                   -7.69
# Warsaw Borough:Urban                   -1.02
# Warsaw Borough:Urban/rural            -3.63
# Jtot                                  -14.70
```

So far, we have used binary weights, so the sum of join-counts multiplied by the weight on that join remains integer. If we change to row standardised weights, where the weights are almost always fractions of 1, the counts, expectations and variances change, but there are few major changes in the z-values.

Using an inverse distance based `listw` object does, however, change the z-values markedly, because closer centroids are up-weighted relatively strongly:

```
Types |> joincount.multi(listw = lw_d183_idw_B)
#                                      Joincount Expected Variance
# Rural:Rural                          3.46e+02 3.61e+02 4.93e+01
# Urban:Urban                          2.90e+01 1.35e+01 2.23e+00
# Urban/rural:Urban/rural              4.65e+01 5.51e+01 9.61e+00
# Warsaw Borough:Warsaw Borough        1.68e+01 4.53e-02 6.61e-03
# Urban:Rural                          2.02e+02 1.40e+02 2.36e+01
# Urban/rural:Rural                    2.25e+02 2.83e+02 3.59e+01
# Urban/rural:Urban                    3.65e+01 5.48e+01 8.86e+00
# Warsaw Borough:Rural                 5.65e+00 8.33e+00 1.73e+00
# Warsaw Borough:Urban                 9.18e+00 1.61e+00 2.54e-01
# Warsaw Borough:Urban/rural           3.27e+00 3.25e+00 5.52e-01
# Jtot                                 4.82e+02 4.91e+02 4.16e+01
#                                      z-value
# Rural:Rural                            -2.10
# Urban:Urban                            10.39
# Urban/rural:Urban/rural                -2.79
# Warsaw Borough:Warsaw Borough         206.38
# Urban:Rural                            12.73
# Urban/rural:Rural                      -9.59
# Urban/rural:Urban                      -6.14
```

```
# Warsaw Borough:Rural              -2.04
# Warsaw Borough:Urban              15.01
# Warsaw Borough:Urban/rural         0.02
# Jtot                              -1.38
```

15.2.2 Moran's *I*

The implementation of Moran's *I* in **spdep** in the `moran.test` function has similar arguments to those of `joincount.test`, but `sampling=` is replaced by `randomisation=` to indicate the underlying analytical approach used for calculating the variance of the measure. It is also possible to use ranks rather than numerical values (Cliff and Ord 1981, 46). The `drop.EI2=` argument may be used to reproduce results where the final component of the variance term is omitted as found in some legacy software implementations.

```
args(moran.test)
```

```
#  function (x, listw, randomisation = TRUE, zero.policy =
#     NULL, alternative = "greater", rank = FALSE, na.action =
#     na.fail, spChk = NULL, adjust.n = TRUE, drop.EI2 = FALSE)
```

The default for the **randomisation=** argument is TRUE, but here we will simply show that the test under normality is the same as a test of least squares residuals with only the intercept used in the mean model. The analysed variable is first-round turnout proportion of registered voters in municipalities and Warsaw boroughs in the 2015 Polish presidential election. The spelling of randomisation is that of Cliff and Ord (1973).

```
pol_pres15 |>
        st_drop_geometry() |>
        subset(select = I_turnout, drop = TRUE) -> I_turnout
```

```
I_turnout |> moran.test(listw = lw_q_B, randomisation = FALSE) |>
      glance_htest()
# Moran I statistic        Expectation             Variance
#         0.691434          -0.000401             0.000140
#        Std deviate            p.value
#         58.461349           0.000000
```

The `lm.morantest` function also takes a **resfun=** argument to set the function used to extract the residuals used for testing, and clearly lets us model other salient features of the response variable (Cliff and Ord 1981, 203). To compare with the standard test, we are only using the intercept here and, as can be seen, the results are the same.

```
lm(I_turnout ~ 1, pol_pres15) |>
    lm.morantest(listw = lw_q_B) |>
    glance_htest()
# Observed Moran I        Expectation          Variance
#          0.691434        -0.000401          0.000140
#       Std deviate          p.value
#         58.461349         0.000000
```

The only difference between tests under normality and randomisation is that an extra term is added if the kurtosis of the variable of interest indicates a flatter or more peaked distribution, where the measure used is the classical measure of kurtosis. Under the default randomisation assumption of analytical randomisation, the results are largely unchanged.

```
(I_turnout |>
    moran.test(listw = lw_q_B) -> mtr) |>
    glance_htest()
# Moran I statistic        Expectation          Variance
#          0.691434        -0.000401          0.000140
#       Std deviate          p.value
#         58.459835         0.000000
```

From the very beginning in the early 1970s, interest was shown in Monte Carlo tests, also known as Hope-type tests and as permutation bootstrap. By default, `moran.mc` returns a `"htest"` object, but may simply use `boot::boot` internally and return a `"boot"` object when `return_boot=TRUE`. In addition the number of simulations needs to be given as `nsim=`; that is the number of times the values of the observations are shuffled at random.

```
set.seed(1)
I_turnout |>
    moran.mc(listw = lw_q_B, nsim = 999,
             return_boot = TRUE) -> mmc
```

The bootstrap permutation retains the outcomes of each of the random permutations, reporting the observed value of the statistic, here Moran's I, the difference between this value and the mean of the simulations under randomisation (equivalent to $E(I)$), and the standard deviation of the simulations under randomisation.

If we compare the Monte Carlo and analytical variances of I under randomisation, we typically see few differences, arguably rendering Monte Carlo testing unnecessary.

```
c("Permutation bootstrap" = var(mmc$t),
  "Analytical randomisation" = unname(mtr$estimate[3]))
#     Permutation bootstrap Analytical randomisation
#                  0.000144                 0.000140
```

Geary's global C is implemented in `geary.test` largely following the same argument structure as `moran.test`. The Getis-Ord G test includes extra arguments to accommodate differences between implementations, as Bivand and Wong (2018) found multiple divergences from the original definitions, often to omit no-neighbour observations generated when using distance band neighbours. It is given by Getis and Ord (1992), on page 194. For G^*, the $i \neq j$ summation constraint is relaxed by including i as a neighbour of itself (thereby also removing the no-neighbour problem, because all observations have at least one neighbour).

Finally, the empirical Bayes Moran's I takes account of the denominator in assessing spatial autocorrelation in rates data (Assunção and Reis 1999). Until now, we have considered the proportion of valid votes cast in relation to the numbers entitled to vote by spatial entity, but using `EBImoran.mc` we can try to accommodate uncertainty in extreme rates in entities with small numbers entitled to vote. There is, however, little impact on the outcome in this case.

Global measures of spatial autocorrelation using spatial weights objects based on graphs of neighbours are, as we have seen, rather blunt tools, which for interpretation depend critically on a reasoned mean model of the variable in question. If the mean model is just the intercept, the global measures will respond to all kinds of misspecification, not only spatial autocorrelation. The choice of entities for aggregation of data will typically be a key source of misspecification.

15.3   Local measures

Building on insights from the weaknesses of global measures, local indicators of spatial association began to appear in the first half of the 1990s (Anselin 1995; Getis and Ord 1992, 1996).

In addition, the Moran plot was introduced, plotting the values of the variable of interest against their spatially lagged values, typically using row-standardised weights to make the axes more directly comparable (Anselin 1996). The `moran.plot` function also returns an influence measures object used to label observations exerting more than proportional influence on the slope of the line representing global Moran's I. In Figure 15.1, we can see that there are many spatial entities exerting such influence. These pairs of observed and lagged observed values make up in aggregate the global measure, but can also be explored in detail. The quadrants of the Moran plot also show low-low pairs in the lower left quadrant, high-high in the upper right quadrant, and fewer low-high and high-low pairs in the upper left and lower right quadrants. In `moran.plot`, the quadrants are split on the means of the variable and its spatial lag; alternative splits are on zero for the centred variable and the spatial lag of the centred variable.

If we extract the hat value influence measure from the returned object, Figure 15.2 suggests that some edge entities exert more than proportional influence (perhaps because of row standardisation), as do entities in or near larger urban areas.

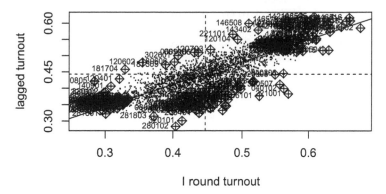

Figure 15.1: Moran plot of I round turnout, row standardised weights

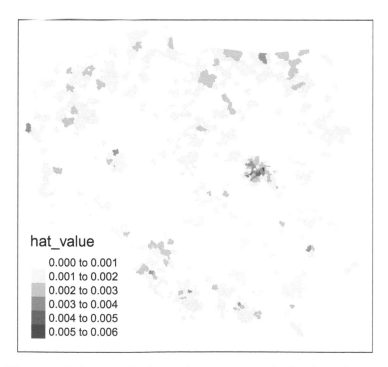

Figure 15.2: Moran plot hat values, row standardised neighbours

15.3.1   Local Moran's I_i

Bivand and Wong (2018) discuss issues impacting the use of local indicators, such as local Moran's I_i and local Getis-Ord G_i. Some issues affect the calculation of the local indicators, others inference from their values. Because n statistics may be calculated from the same number of observations, there are multiple comparison problems that need to be addressed. Caldas de Castro and Singer (2006) conclude, based on a typical dataset and a simulation exercise, that the false discovery rate (FDR) adjustment of probability values will certainly give a better picture of interesting clusters than no adjustment. Following this up, Anselin (2019) explores the combination of FDR adjustments with the use of redefined "significance" cutoffs (Benjamin et al. 2018), for example 0.01, 0.005, and 0.001 instead of 0.1, 0.05, and 0.01; the use of the term *interesting* rather than *significant* is also preferred. This is discussed further in Bivand (2022b). As in the global case, misspecification remains a source of confusion, and, further, interpreting local spatial autocorrelation in the presence of global spatial autocorrelation is challenging (Ord and Getis 2001; Tiefelsdorf 2002; Bivand, Müller, and Reder 2009).

```
args(localmoran)
```

```
#  function (x, listw, zero.policy = NULL, na.action = na.fail,
#    conditional = TRUE, alternative = "two.sided", mlvar =
#    TRUE, spChk = NULL, adjust.x = FALSE)
```

In an important clarification, Sauer et al. (2021) show that the comparison of standard deviates for local Moran's I_i based on analytical formulae and conditional permutation in Bivand and Wong (2018) was based on a misunderstanding. Sokal, Oden, and Thomson (1998) provide alternative analytical formulae for standard deviates of local Moran's I_i based either on total or conditional permutation, but the analytical formulae used in Bivand and Wong (2018), based on earlier practice, only use total permutation, and consequently do not match the simulation conditional permutations. Thanks to a timely pull request, **localmoran** now has a **conditional=** argument (default **TRUE**) using alternative formulae from the appendix of Sokal, Oden, and Thomson (1998). The **mlvar=** and **adjust.x=** arguments to **localmoran** are discussed in Bivand and Wong (2018), and permit matching with other implementations. Taking **"two.sided"** probability values (the default), we obtain:

```
I_turnout |>
    localmoran(listw = lw_q_W) -> locm
```

The I_i local indicators when summed and divided by the sum of the spatial weights equal global Moran's I, showing the possible presence of positive and negative local spatial autocorrelation:

```
all.equal(sum(locm[,1])/Szero(lw_q_W),
          unname(moran.test(I_turnout, lw_q_W)$estimate[1]))
# [1] TRUE
```

Using `stats::p.adjust` to adjust for multiple comparisons, we see that over 15% of the 2495 local measures have p-values < 0.005 if no adjustment is applied, but only 1.5% using Bonferroni adjustment to control the family-wise error rate, with two other choices shown: `"fdr"` is the Benjamini and Hochberg (1995) false discovery rate (almost 6%) and `"BY"` (Benjamini and Yekutieli 2001), another false discovery rate adjustment (about 2.5%):

```
pva <- function(pv) cbind("none" = pv,
    "FDR" = p.adjust(pv, "fdr"), "BY" = p.adjust(pv, "BY"),
    "Bonferroni" = p.adjust(pv, "bonferroni"))
locm |>
    subset(select = "Pr(z != E(Ii))", drop = TRUE) |>
    pva() -> pvsp
f <- function(x) sum(x < 0.005)
apply(pvsp, 2, f)
#       none        FDR        BY Bonferroni
#        385        149        64         38
```

In the global measure case, bootstrap permutations may be used as an alternative to analytical methods for possible inference, where both the theoretical development of the analytical variance of the measure, and the permutation scheme, shuffle all of the observed values. In the local case, conditional permutation should be used, fixing the value at observation i and randomly sampling from the remaining $n-1$ values to find randomised values at neighbours. Conditional permutation is provided as function `localmoran_perm`, which may use multiple compute nodes to sample in parallel if provided, and permits the setting of a seed for the random number generator across the compute nodes. The number of simulations `nsim=` also controls the precision of the ranked estimates of the probability value based on the rank of observed I_i among the simulated values:

```
library(parallel)
invisible(spdep::set.coresOption(max(detectCores()-1L, 1L)))
I_turnout |>
    localmoran_perm(listw = lw_q_W, nsim = 9999,
                    iseed = 1) -> locm_p
```

The outcome is that over 15% of observations have two sided p-values < 0.005 without multiple comparison adjustment, and about 1.5% with Bonferroni adjustment, when the p-values are calculated using the standard deviate of the permutation samples and the normal distribution.

```
locm_p |>
    subset(select = "Pr(z != E(Ii))", drop = TRUE) |>
    pva() -> pvsp
apply(pvsp, 2, f)
#       none        FDR        BY Bonferroni
#        379        149        63         40
```

Since the variable under analysis may not be normally distributed, the p-values can also be calculated by finding the rank of the observed I_i among the rank-based simulated values, and looking up the probability value from the uniform distribution taking the **alternative=** choice into account:

```
locm_p |>
    subset(select = "Pr(z != E(Ii)) Sim", drop = TRUE) |>
    pva() -> pvsp
apply(pvsp, 2, f)
#        none       FDR      BY Bonferroni
#         394       125       0         0
```

Now the **"BY"** and Bonferroni counts of *interesting* locations are zero with 9999 samples, but may be recovered by increasing the sample count to 999999 if required; the FDR adjustment and *interesting* cutoff 0.005 yields about 5% locations.

```
pol_pres15$locm_pv <- p.adjust(locm[, "Pr(z != E(Ii))"], "fdr")
pol_pres15$locm_std_pv <- p.adjust(locm_p[, "Pr(z != E(Ii))"],
                          "fdr")
pol_pres15$locm_p_pv <- p.adjust(locm_p[, "Pr(z != E(Ii)) Sim"],
                        "fdr")
```

Proceeding using the FDR adjustment and an *interesting* location cutoff of 0.005, we can see from Figure 15.3 that the adjusted probability values for the analytical conditional approach, the approach using the moments of the sampled values from permutation sampling, and the approach using the ranks of observed values among permutation samples all yield similar maps, as the distribution of the input variable is quite close to normal.

In presenting local Moran's I, use is often made of "hotspot" maps. Because I_i takes high values both for strong positive autocorrelation of low and high values of the input variable, it is hard to show where "clusters" of similar neighbours with low or high values of the input variable occur. The quadrants of the Moran plot are used, by creating a categorical quadrant variable interacting the input variable and its spatial lag split at their means. The quadrant categories are then set to NA if, for the chosen probability value and adjustment, I_i would not be considered *interesting*. Here, for the FDR adjusted conditional analytical probability values (Figure 15.3, upper left panel), 53 observations belong to **"Low-Low"** cluster cores, and 96 to **"High-High"** cluster cores, similarly for the standard deviate-based permutation p-values (Figure 15.3, upper right panel), but the rank-based permutation p-values reduce the **"High-High"** count and increase the **"Low-Low"** count Figure 15.3 lower left panel:

```
quadr <- attr(locm, "quadr")$mean
a <- table(addNA(quadr))
locm |> hotspot(Prname="Pr(z != E(Ii))", cutoff = 0.005,
               droplevels=FALSE) -> pol_pres15$hs_an_q
```

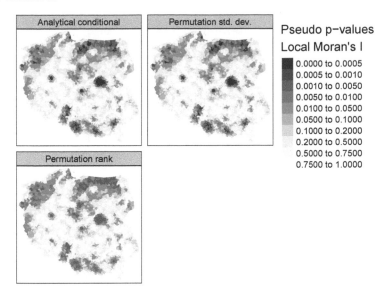

Figure 15.3: Local Moran's I FDR probability values: left upper panel: analytical conditional p-values; right upper panel: permutation standard deviate conditional p-values; left lower panel: permutation rank conditional p-values, first-round turnout, row-standardised neighbours

```
locm_p |> hotspot(Prname="Pr(z != E(Ii))", cutoff = 0.005,
                  droplevels=FALSE) -> pol_pres15$hs_ac_q
locm_p |> hotspot(Prname="Pr(z != E(Ii)) Sim", cutoff = 0.005,
                  droplevels = FALSE) -> pol_pres15$hs_cp_q
b <- table(addNA(pol_pres15$hs_an_q))
c <- table(addNA(pol_pres15$hs_ac_q))
d <- table(addNA(pol_pres15$hs_cp_q))
t(rbind("Moran plot quadrants" = a, "Analytical cond." = b,
  "Permutation std. cond." = c, "Permutation rank cond." = d))
#           Moran plot quadrants Analytical cond.
# Low-Low                   1040               53
# High-Low                   264                0
# Low-High                   213                0
# High-High                  978               96
# <NA>                         0             2346
#           Permutation std. cond. Permutation rank cond.
# Low-Low                       53                     55
# High-Low                       0                      0
# Low-High                       0                      0
# High-High                     96                     70
# <NA>                        2346                   2370
```

```
pol_pres15$hs_an_q <- droplevels(pol_pres15$hs_an_q)
pol_pres15$hs_ac_q <- droplevels(pol_pres15$hs_ac_q)
pol_pres15$hs_cp_q <- droplevels(pol_pres15$hs_cp_q)
```

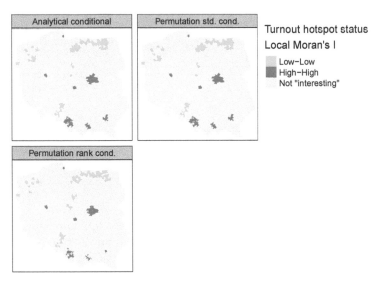

Figure 15.4: Local Moran's I FDR hotspot cluster core maps $\alpha = 0.005$: left upper panel: analytical conditional p-values; right upper panel: permutation standard deviate conditional p-values; left lower panel: permutation rank conditional p-values, first-round turnout, row-standardised neighbours

Figure 15.4 shows that there is very little difference between the FDR-adjusted *interesting* clusters with a choice of an $\alpha = 0.005$ probability value cutoff for the three approaches of analytical conditional standard deviates, permutation-based standard deviates, and rank-based probability values; the "High-High" cluster cores are metropolitan areas.

Tiefelsdorf (2002) argues that standard approaches to the calculation of the standard deviates of local Moran's I_i should be supplemented by numerical estimates, and shows that saddlepoint approximations are a computationally efficient way of achieving this goal. The localmoran.sad function takes a fitted linear model as its first argument, so we first fit a null (intercept only) model, but use case weights because the numbers entitled to vote vary greatly between observations:

```
lm(I_turnout ~ 1) -> lm_null
```

Saddlepoint approximation is as computationally intensive as conditional permutation, because, rather than computing a simple measure on many samples, a good deal of numerical calculation is needed for each local approximation:

```
lm_null |> localmoran.sad(nb = nb_q, style = "W",
                              alternative = "two.sided") |>
      summary() -> locm_sad_null
```

The chief advantage of the saddlepoint approximation is that it takes a fitted linear model rather than simply a numerical variable, so the residuals are analysed. With an intercept-only model, the results are similar to local Moran's I_i, but we can weight the observations, here by the count of those entitled to vote, which should down-weight small units of observation:

```
lm(I_turnout ~ 1, weights = pol_pres15$I_entitled_to_vote) ->
        lm_null_weights
lm_null_weights |>
            localmoran.sad(nb = nb_q, style = "W",
                              alternative = "two.sided") |>
        summary() -> locm_sad_null_weights
```

Next we add the categorical variable distinguishing between rural, urban and other types of observational unit:

```
lm(I_turnout ~ Types, weights=pol_pres15$I_entitled_to_vote) ->
        lm_types
lm_types |> localmoran.sad(nb = nb_q, style = "W",
                              alternative = "two.sided") |>
        summary() -> locm_sad_types
```

```
locm_sad_null |> hotspot(Prname="Pr. (Sad)",
                    cutoff=0.005) -> pol_pres15$locm_sad0
locm_sad_null_weights |> hotspot(Prname="Pr. (Sad)",
                    cutoff = 0.005) -> pol_pres15$locm_sad1
locm_sad_types |> hotspot(Prname="Pr. (Sad)",
                    cutoff = 0.005) -> pol_pres15$locm_sad2
```

```
rbind(null = append(table(addNA(pol_pres15$locm_sad0)),
                c("Low-High" = 0), 1),
      weighted = append(table(addNA(pol_pres15$locm_sad1)),
                c("Low-High" = 0), 1),
      type_weighted = append(table(addNA(pol_pres15$locm_sad2)),
                c("Low-High" = 0), 1))
#               Low-Low Low-High High-High <NA>
# null               19        0        55 2421
# weighted            9        0        52 2434
# type_weighted      13        0        81 2401
```

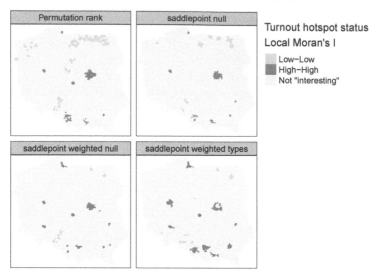

Figure 15.5: Local Moran's I FDR hotspot cluster core maps, two-sided, *interesting* cutoff $\alpha = 0.005$: left upper panel: permutation rank conditional p-values; right upper panel: null (intercept only) model saddlepoint p-values; left lower panel: weighted null (intercept only) model saddlepoint p-values; right lower panel: weighted types model saddlepoint p-values, for first-round turnout, row-standardised neighbours

Figure 15.5 includes the permutation rank cluster cores for comparison (upper left panel). Because saddlepoint approximation permits richer mean models to be used, and possibly because the approximation approach is inherently local, relating regression residual values at i to those of its neighbours, the remaining three panels diverge somewhat. The intercept-only (null) model is fairly similar to standard local Moran's I_i, but weighting by counts of eligible voters removes most of the "Low-Low" cluster cores. Adding the type categorical variable strengthens the urban "High-High" cluster cores but removes the Warsaw boroughs as *interesting* cluster cores. The central boroughs are surrounded by other boroughs, all with high turnout, not driven by autocorrelation but by being metropolitan boroughs. It is also possible to use saddlepoint approximation where the global spatial process has been incorporated, removing the conflation of global and local spatial autocorrelation in standard approaches.

The same can also be accomplished using exact methods, but may require more tuning as numerical integration may fail, returning NaN rather than the exact estimate of the standard deviate (Bivand, Müller, and Reder 2009):

```
lm_types |> localmoran.exact(nb = nb_q, style = "W",
    alternative = "two.sided", useTP=TRUE, truncErr=1e-8) |>
    as.data.frame() -> locm_ex_types
```

```
locm_ex_types |> hotspot(Prname = "Pr. (exact)",
                         cutoff = 0.005) -> pol_pres15$locm_ex
```

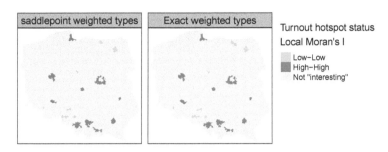

Figure 15.6: Local Moran's I FDR hotspot cluster core maps, two-sided, *interesting* cutoff $\alpha = 0.005$: left panel: weighted types model saddlepoint p-values; right panel: weighted types model exact p-values, for first-round turnout, row-standardised neighbours

As Figure 15.6 shows, the exact and saddlepoint approximation methods yield almost identical cluster classifications from the same regression residuals, multiple comparison adjustment method, and cutoff level, with the exact method returning four more *interesting* observations:

```
table(Saddlepoint = addNA(pol_pres15$locm_sad2),
      exact = addNA(pol_pres15$locm_ex))
#                 exact
# Saddlepoint Low-Low High-High <NA>
#   Low-Low         13         0    0
#   High-High        0        81    0
#   <NA>             2         2 2397
```

15.3.2   Local Getis-Ord G_i

The local Getis-Ord G_i measure (Getis and Ord 1992, 1996) is reported as a standard deviate, and, may also take the G_i^* form where self-neighbours are inserted into the neighbour object using **include.self**. The observed and expected values of local G with their analytical variances may also be returned if **return_internals=TRUE**.

```
I_turnout |>
        localG(lw_q_W, return_internals = TRUE) -> locG
```

Permutation inference is also available for this measure:

```
I_turnout |>
        localG_perm(lw_q_W, nsim = 9999, iseed = 1) -> locG_p
```

The correlation between the two-sided probability values for analytical and permutation-based standard deviates (first two columns and rows) and permutation rank-based probability values are very strong:

```
cor(cbind(localG=attr(locG, "internals")[, "Pr(z != E(Gi))"],
    attr(locG_p, "internals")[, c("Pr(z != E(Gi))",
                                  "Pr(z != E(Gi)) Sim")]))
#                          localG Pr(z != E(Gi)) Pr(z != E(Gi)) Sim
# localG                        1              1                  1
# Pr(z != E(Gi))                1              1                  1
# Pr(z != E(Gi)) Sim            1              1                  1
```

15.3.3   Local Geary's C_i

Anselin (2019) extends Anselin (1995) and has been recently added to **spdep** thanks to contributions by Josiah Parry (pull request https://github.com/r-spatial/spdep/pull/66). The conditional permutation framework used for I_i and G_i is also used for C_i:

```
I_turnout |>
        localC_perm(lw_q_W, nsim=9999, iseed=1) -> locC_p
```

The permutation standard deviate-based and rank-based probability values are not as highly correlated as for G_i, in part reflecting the difference in view of autocorrelation in C_i as represented by a function of the differences between values rather than the products of values:

```
cor(attr(locC_p, "pseudo-p")[, c("Pr(z != E(Ci))",
                                  "Pr(z != E(Ci)) Sim")])
#                        Pr(z != E(Ci)) Pr(z != E(Ci)) Sim
# Pr(z != E(Ci))                  1.000                0.966
# Pr(z != E(Ci)) Sim              0.966                1.000
```

```
locC_p |> hotspot(Prname = "Pr(z != E(Ci)) Sim",
                  cutoff = 0.005) -> pol_pres15$hs_C
locG_p |> hotspot(Prname = "Pr(z != E(Gi)) Sim",
                  cutoff = 0.005) -> pol_pres15$hs_G
```

Figure 15.7 shows that the cluster cores identified as *interesting* using I_i, G_i and C_i for the same variable, first-round turnout, and the same spatial weights, for rank-based permutation FDR adjusted probability values and an $\alpha = 0.005$ cutoff, are very similar. In most cases, the **"High-High"** cluster cores are urban areas, and **"Low-Low"** cores are sparsely populated rural areas in the North, in addition to the German national minority areas close to the southern border. The three measures use slightly different strategies for naming cluster cores: I_i uses quadrants of the Moran scatterplot, G_i splits into **"Low"** and **"High"** on the mean of the input variable

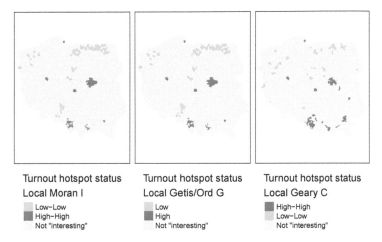

Turnout hotspot status
Local Moran I

- Low–Low
- High–High
- Not "interesting"

Turnout hotspot status
Local Getis/Ord G

- Low
- High
- Not "interesting"

Turnout hotspot status
Local Geary C

- High–High
- Low–Low
- Not "interesting"

Figure 15.7: FDR hotspot cluster core maps, two-sided, *interesting* cutoff $\alpha = 0.005$: left panel: local Moran's I_i; centre panel: local Getis-Ord G_i; right panel: local Geary's C_i; first-round turnout, row-standardised neighbours

(which is the same as the first component in the I_i tuple), and univariate C_i on the mean of the input variable and zero for its lag. As before, cluster categories that do not occur are dropped.

For comparison, and before moving to multivariate C_i, let us take the univariate C_i for the second (final) round turnout. One would expect that the run-off between the two top candidates from the first-round might mobilise some voters who did not have a clear first-round preference, but that it discourages some of those with strong loyalty to a candidate eliminated after the first round:

```
pol_pres15 |>
        st_drop_geometry() |>
        subset(select = II_turnout) |>
        localC_perm(lw_q_W, nsim=9999, iseed=1) -> locC_p_II
```

```
locC_p_II |> hotspot(Prname = "Pr(z != E(Ci)) Sim",
                    cutoff = 0.005) -> pol_pres15$hs_C_II
```

Multivariate C_i (Anselin 2019) is taken as the sum of univariate C_i divided by the number of variables, but permutation is fixed so that the correlation between the variables is unchanged:

```
pol_pres15 |>
        st_drop_geometry() |>
        subset(select = c(I_turnout, II_turnout)) |>
        localC_perm(lw_q_W, nsim=9999, iseed=1) -> locMvC_p
```

Let us check that the multivariate C_i is equal to the mean of the univariate C_i:

```
all.equal(locMvC_p, (locC_p+locC_p_II)/2,
          check.attributes = FALSE)
# [1] TRUE
```

```
locMvC_p |> hotspot(Prname = "Pr(z != E(Ci)) Sim",
                    cutoff = 0.005) -> pol_pres15$hs_MvC
```

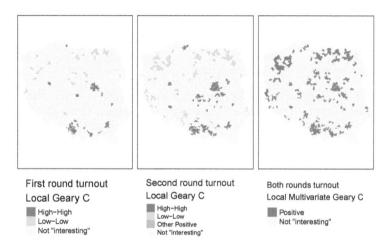

First round turnout
Local Geary C
▪ High–High
  Low–Low
  Not "interesting"

Second round turnout
Local Geary C
▪ High–High
  Low–Low
  Other Positive
  Not "Interesting"

Both rounds turnout
Local Multivariate Geary C
▪ Positive
  Not "interesting"

Figure 15.8: FDR hotspot cluster core maps, two-sided, *interesting* cutoff $\alpha = 0.005$: left panel: local C_i, first-round turnout; centre panel: local C_i, second-round turnout; right panel: local multivariate C_i, both turnout rounds; row-standardised neighbours

Figure 15.8 indicates that the multivariate measure picks up aggregated elements of observations found *interesting* in the two univariate measures. We can break this down by interacting the first- and second-round univariate measures, and tabulating against the multivariate measure.

```
table(droplevels(interaction(addNA(pol_pres15$hs_C),
                             addNA(pol_pres15$hs_C_II), sep=":")),
      addNA(pol_pres15$hs_MvC))
#
#                        Positive <NA>
#    High-High:High-High       74    0
#    NA:High-High              55   16
#    Low-Low:Low-Low           24    0
#    NA:Low-Low                53   11
#    NA:Other Positive          1    0
#    High-High:NA              13    4
```

```
#     Low-Low:NA                    8    3
#     NA:NA                        73 2160
```

For these permutation outcomes, 47 observations in the multivariate case are found *interesting* where neither of the univariate C_i were found *interesting* (FDR, cutoff 0.005). Almost all of the observations found *interesting* in both first and second round, are also interesting in the multivariate case, but outcomes are more mixed when observations were only found interesting in one of the two rounds.

15.3.4   The rgeoda package

Geoda has been wrapped for R as **rgeoda** (Li and Anselin 2022), and provides very similar functionalities for the exploration of spatial autocorrelation in areal data as matching parts of **spdep**. The active objects are kept as pointers to a compiled code workspace; using compiled code for all operations (as in Geoda itself) makes **rgeoda** perform fast, but makes it less flexible when modifications or enhancements are desired.

```
library(rgeoda)
Geoda_w <- queen_weights(pol_pres15)
summary(Geoda_w)
#                         name                value
# 1 number of observations:                   2495
# 2            is symmetric:                   TRUE
# 3                sparsity: 0.00228786229774178
# 4       # min neighbors:                       1
# 5       # max neighbors:                      13
# 6       # mean neighbors:      5.70821643286573
# 7     # median neighbors:                       6
# 8            has isolates:                  FALSE
```

For comparison, let us take the multivariate C_i measure of turnout in the two rounds of the 2015 Polish presidential election as above:

```
lisa <- local_multigeary(Geoda_w,
    pol_pres15[c("I_turnout", "II_turnout")],
    cpu_threads = max(detectCores() - 1, 1),
    permutations = 99999, seed = 1)
```

The contiguity neighbours are the same as those found by **poly2nb**:

```
all.equal(card(nb_q), lisa_num_nbrs(lisa),
        check.attributes = FALSE)
# [1] TRUE
```

as are the multivariate C_i values the same as those found above:

```
all.equal(lisa_values(lisa), c(locMvC_p),
          check.attributes = FALSE)
# [1] TRUE
```

One difference is that the range of the folded two-sided rank-based permutation probability values used by **rgeoda** is $[0, 0.5]$, also reported in **spdep**:

```
apply(attr(locMvC_p, "pseudo-p")[,c("Pr(z != E(Ci)) Sim",
                                    "Pr(folded) Sim")], 2, range)
#       Pr(z != E(Ci)) Sim Pr(folded) Sim
# [1,]             0.0002         0.0001
# [2,]             0.9988         0.4994
```

This means that the cutoff corresponding to 0.005 over $[0, 1]$ is 0.0025 over $[0, 0.5]$:

```
locMvC_p |> hotspot(Prname = "Pr(folded) Sim",
                    cutoff = 0.0025) -> pol_pres15$hs_MvCa
```

So although `local_multigeary` used the default cutoff of 0.05 in setting cluster core classes, we can sharpen the cutoff and apply the FDR adjustment on output components of the `lisa` object in the compiled code workspace:

```
mvc <- factor(lisa_clusters(lisa), levels=0:2,
              labels = lisa_labels(lisa)[1:3])
is.na(mvc) <- p.adjust(lisa_pvalues(lisa), "fdr") >= 0.0025
pol_pres15$geoda_mvc <- droplevels(mvc)
```

About 80 more observations are found *interesting* in the **rgeoda** permutation, and further analysis of implementation details is still in progress:

```
addmargins(table(spdep = addNA(pol_pres15$hs_MvCa),
                 rgeoda = addNA(pol_pres15$geoda_mvc)))
#              rgeoda
# spdep     Positive <NA>  Sum
#   Positive     292    9  301
#   <NA>          35 2159 2194
#   Sum          327 2168 2495
```

Figure 15.9 shows that while almost all of the 242 observations found *interesting* in the **spdep** implementation were also *interesting* for **rgeoda**, the latter found a further 86 *interesting*. Of course, permutation outcomes are bound to vary, but it remains to establish whether either or both implementations require revision.

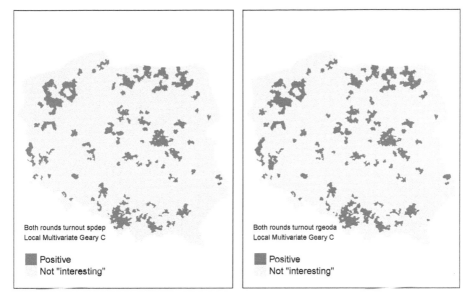

Figure 15.9: FDR local multivariate C_i hotspot cluster core maps, two-sided, *interesting* cutoff $\alpha = 0.0025$ over $[0, 0.5]$: left panel: **spdep**, both turnout rounds; right panel: **rgeoda**, both turnout rounds; row-standardised neighbours

15.4 Exercises

1. Why are join-count measures on a chessboard so different between **rook** and **queen** neighbours?
2. Please repeat the simulation shown in Section 15.1 using the chessboard polygons and the row-standardised **queen** contiguity neighbours. Why is it important to understand that spatial autocorrelation usually signals (unavoidable) misspecification in our data?
3. Why is false discovery rate adjustment recommended for local measures of spatial autocorrelation?
4. Compare the local Moran's I_i standard deviate values for the simulated data from exercise 2 (above) for the analytical conditional approach, and saddlepoint approximation. Consider the advantages and disadvantages of the saddlepoint approximation approach.

16

Spatial Regression

Even though it may be tempting to focus on interpreting the map pattern of an areal support response variable of interest, the pattern may largely derive from covariates (and their functional forms), as well as the respective spatial footprints of the variables in play. Spatial autoregressive models in two dimensions began without covariates and with clear links to time series (Whittle 1954). Extensions included tests for spatial autocorrelation in linear model residuals, and models applying the autoregressive component to the response or the residuals, where the latter matched the tests for residuals (Cliff and Ord 1972, 1973). These "lattice" models of areal data typically express the dependence between observations using a graph of neighbours in the form of a contiguity matrix.

Of course, handling a spatial correlation structure in a generalised least squares model or a (generalised) linear or non-linear mixed effects model such as those provided in the **nlme** and many other packages does not have to use a graph of neighbours (Pinheiro and Bates 2000). These models are also spatial regression models, using functions of the distance between observations, and fitted variograms to model the spatial autocorrelation present; such models have been held to yield a clearer picture of the underlying processes (Wall 2004), building on geostatistics. For example, the **glmmTMB** package successfully uses this approach to spatial regression (Brooks et al. 2017). Here we will only consider spatial regression using spatial weights matrices.

16.1 Markov random field and multilevel models

There is a large literature in disease mapping using conditional autoregressive (CAR) and intrinsic CAR (ICAR) models in spatially structured random effects. These extend to multilevel models, in which the spatially structured random effects may apply at different levels of the model (Bivand et al. 2017). In order to try out some of the variants, we need to remove the no-neighbour observations from the tract level, and from the model output zone aggregated level, in two steps as reducing the tract level induces a no-neighbour outcome at the model output zone level. Many of the model estimating functions take `family=` arguments, and fit generalised linear mixed effects models with per-observation spatial random effects structured using a Markov random field representation of relationships between neighbours. In the multilevel case, the random effects may be modelled at the group level, which is the case presented in the following examples.

We follow Gómez-Rubio (2019) in summarising Pinheiro and Bates (2000) and McCulloch and Searle (2001) to describe the mixed-effects model representation of spatial regression models. In a Gaussian linear mixed model setting, a random effect u is added to the model, with response Y, fixed covariates X, their coefficients β and error term $\varepsilon_i \sim N(0, \sigma^2), i = 1, \ldots, n$:

$$Y = X\beta + Zu + \varepsilon$$

Z is a fixed design matrix for the random effects. If there are n random effects, it will be an $n \times n$ identity matrix if instead the observations are aggregated into m groups, so with $m < n$ random effects, it will be an $n \times m$ matrix showing which group each observation belongs to. The random effects are modelled as a multivariate Normal distribution $u \sim N(0, \sigma_u^2 \Sigma)$, and $\sigma_u^2 \Sigma$ is the square variance-covariance matrix of the random effects.

A division has grown up, possibly unhelpfully, between scientific fields using CAR models (Besag 1974), and simultaneous autoregressive models (SAR) (Ord 1975; Hepple 1976). Although CAR and SAR models are closely related, these fields have found it difficult to share experience of applying similar models, often despite referring to key work summarising the models (Ripley 1981, 1988; Cressie 1993). Ripley gives the SAR variance as (Ripley 1981, 89), here shown as the inverse Σ^{-1} (also known as the precision matrix):

$$\Sigma^{-1} = [(I - \rho W)'(I - \rho W)]$$

where ρ is a spatial autocorrelation parameter and W is a non-singular spatial weights matrix that represents spatial dependence. The CAR variance is:

$$\Sigma^{-1} = (I - \rho W)$$

where W is a symmetric and strictly positive definite spatial weights matrix. In the case of the intrinsic CAR model, avoiding the estimation of a spatial autocorrelation parameter, we have:

$$\Sigma^{-1} = M = \operatorname{diag}(n_i) - W$$

where W is a symmetric and strictly positive definite spatial weights matrix as before and n_i are the row sums of W. The Besag-York-Mollié model includes intrinsic CAR spatially structured random effects and unstructured random effects. The Leroux model combines matrix components for unstructured and spatially structured random effects, where the spatially structured random effects are taken as following an intrinsic CAR specification:

$$\Sigma^{-1} = [(1 - \rho)I_n + \rho M]$$

References to the definitions of these models may be found in Gómez-Rubio (2020), and estimation issues affecting the Besag-York-Mollié and Leroux models are reviewed by Gerber and Furrer (2015).

More recent books expounding the theoretical bases for modelling with areal data simply point out the similarities between SAR and CAR models in relevant chapters (Gaetan and Guyon 2010; Van Lieshout 2019); the interested reader is invited to consult these sources for background information.

16.1.1   Boston house value dataset

Here we shall use the Boston housing dataset, which has been restructured and furnished with census tract boundaries (Bivand 2017). The original dataset used 506 census tracts and a hedonic model to try to estimate willingness to pay for clean air. The response was constructed from counts of ordinal answers to a 1970 census question about house value. The response is left- and right-censored in the census source and has been treated as Gaussian. The key covariate was created from a calibrated meteorological model showing the annual nitrogen oxides (NOX) level for a smaller number of model output zones. The numbers of houses responding also varies by tract and model output zone. There are several other covariates, some measured at the tract level, some by town only, where towns broadly correspond to the air pollution model output zones.

We can start by reading in the 506 tract dataset from **spData** (Bivand, Nowosad, and Lovelace 2022), and creating a contiguity neighbour object and from that again a row standardised spatial weights object.

```
library(sf)
library(spData)
boston_506 <- st_read(system.file("shapes/boston_tracts.shp",
                      package = "spData")[1], quiet = TRUE)


nb_q <- spdep::poly2nb(boston_506)
lw_q <- spdep::nb2listw(nb_q, style = "W")
```

If we examine the median house values, we find that those for censored values have been assigned as missing, and that 17 tracts are affected.

```
table(boston_506$censored)
#
#  left     no right
#     2    489    15


summary(boston_506$median)
#    Min. 1st Qu.  Median    Mean 3rd Qu.    Max.    NA's
#    5600   16800   21000   21749   24700   50000      17
```

Next, we can subset to the remaining 489 tracts with non-censored house values, and the neighbour object to match. The neighbour object now has one observation with no neighbours.

```
boston_506$CHAS <- as.factor(boston_506$CHAS)
boston_489 <- boston_506[!is.na(boston_506$median),]
nb_q_489 <- spdep::poly2nb(boston_489)
```

```
lw_q_489 <- spdep::nb2listw(nb_q_489, style = "W",
                            zero.policy = TRUE)
```

The NOX_ID variable specifies the upper-level aggregation, letting us aggregate the tracts to air pollution model output zones. We can create aggregate neighbour and row standardised spatial weights objects, and aggregate the NOX variable taking means, and the CHAS Charles River dummy variable for observations on the river. Here we follow the principles outlined in Section 5.3.1 for spatially extensive and intensive variables; neither NOX nor CHAS can be summed, as they are not count variables.

```
agg_96 <- list(as.character(boston_506$NOX_ID))
boston_96 <- aggregate(boston_506[, "NOX_ID"], by = agg_96,
                       unique)
nb_q_96 <- spdep::poly2nb(boston_96)
lw_q_96 <- spdep::nb2listw(nb_q_96)
boston_96$NOX <- aggregate(boston_506$NOX, agg_96, mean)$x
boston_96$CHAS <-
    aggregate(as.integer(boston_506$CHAS)-1, agg_96, max)$x
```

The response is aggregated using the weightedMedian function in **matrixStats**, and midpoint values for the house value classes. Counts of houses by value class were punched to check the published census values, which can be replicated using weightedMedian at the tract level. Here we find two output zones with calculated weighted medians over the upper census question limit of USD $50,000, and remove them subsequently as they also are affected by not knowing the appropriate value to insert for the top class by value. This is a case of spatially extensive aggregation, for which the summation of counts is appropriate:

```
nms <- names(boston_506)
ccounts <- 23:31
for (nm in nms[c(22, ccounts, 36)]) {
  boston_96[[nm]] <- aggregate(boston_506[[nm]], agg_96, sum)$x
}
br2 <-
  c(3.50, 6.25, 8.75, 12.5, 17.5, 22.5, 30, 42.5, 60) * 1000
counts <- as.data.frame(boston_96)[, nms[ccounts]]
f <- function(x) matrixStats::weightedMedian(x = br2, w = x,
                                             interpolate = TRUE)
boston_96$median <- apply(counts, 1, f)
is.na(boston_96$median) <- boston_96$median > 50000
summary(boston_96$median)
#    Min. 1st Qu.  Median    Mean 3rd Qu.    Max.    NA's
#    9009   20417   23523   25263   30073   49496       2
```

Before subsetting, we aggregate the remaining covariates by weighted mean using the tract population counts punched from the census (Bivand 2017); these are spatially intensive variables, not count data.

```
boston_94 <- boston_96[!is.na(boston_96$median),]
nb_q_94 <- spdep::subset.nb(nb_q_96, !is.na(boston_96$median))
lw_q_94 <- spdep::nb2listw(nb_q_94, style="W")
```

We now have two datasets at each level, at the lower, census tract level, and at the upper, air pollution model output zone level, one including the censored observations, the other excluding them.

```
boston_94a <- aggregate(boston_489[,"NOX_ID"],
                        list(boston_489$NOX_ID), unique)
nb_q_94a <- spdep::poly2nb(boston_94a)
NOX_ID_no_neighs <-
        boston_94a$NOX_ID[which(spdep::card(nb_q_94a) == 0)]
boston_487 <- boston_489[is.na(match(boston_489$NOX_ID,
                                     NOX_ID_no_neighs)),]
boston_93 <- aggregate(boston_487[, "NOX_ID"],
                       list(ids = boston_487$NOX_ID), unique)
row.names(boston_93) <- as.character(boston_93$NOX_ID)
nb_q_93 <- spdep::poly2nb(boston_93,
        row.names = unique(as.character(boston_93$NOX_ID)))
```

The original model related the log of median house values by tract to the square of NOX values, including other covariates usually related to house value by tract, such as aggregate room counts, aggregate age, ethnicity, social status, distance to downtown and to the nearest radial road, a crime rate, and town-level variables reflecting land use (zoning, industry), taxation and education (Bivand 2017). This structure will be used here to exercise issues raised in fitting spatial regression models, including the presence of multiple levels.

16.2   Multilevel models of the Boston dataset

The ZN, INDUS, NOX, RAD, TAX, and PTRATIO variables show effectively no variability within the TASSIM zones, so in a multilevel model the random effect may absorb their influence.

```
form <- formula(log(median) ~ CRIM + ZN + INDUS + CHAS +
                I((NOX*10)^2) + I(RM^2) + AGE + log(DIS) +
                log(RAD) + TAX + PTRATIO + I(BB/100) +
                log(I(LSTAT/100)))
```

16.2.1   IID random effects with lme4

The **lme4** package (Bates et al. 2022) lets us add an independent and identically distributed (IID) unstructured random effect at the model output zone level by updating the model formula with a random effects term:

```
library(Matrix)
library(lme4)
MLM <- lmer(update(form, . ~ . + (1 | NOX_ID)),
            data = boston_487, REML = FALSE)
```

Copying the random effect into the "sf" object for mapping is performed below.

```
boston_93$MLM_re <- ranef(MLM)[[1]][,1]
```

16.2.2   IID and CAR random effects with hglm

The same model may be estimated using the **hglm** package (Alam, Ronnegard, and Shen 2019), which also permits the modelling of discrete responses, this time using an extra one-sided formula to express the random effects term:

```
library(hglm) |> suppressPackageStartupMessages()
suppressWarnings(HGLM_iid <- hglm(fixed = form,
                                  random = ~1 | NOX_ID,
                                  data = boston_487,
                                  family = gaussian()))
boston_93$HGLM_re <- unname(HGLM_iid$ranef)
```

The same package has been extended to spatially structured SAR and CAR random effects, for which a sparse spatial weights matrix is required (Alam, Rönnegård, and Shen 2015); we choose binary spatial weights:

```
library(spatialreg)
W <- as(spdep::nb2listw(nb_q_93, style = "B"), "CsparseMatrix")
```

We fit a CAR model at the upper level, using the **rand.family=** argument, where the values of the indexing variable NOX_ID match the row names of W:

```
suppressWarnings(HGLM_car <- hglm(fixed = form,
                                  random = ~ 1 | NOX_ID,
                                  data = boston_487,
                                  family = gaussian(),
                                  rand.family = CAR(D=W)))
boston_93$HGLM_ss <- HGLM_car$ranef[,1]
```

16.2.3   IID and ICAR random effects with R2BayesX

The **R2BayesX** package (Umlauf et al. 2022) provides flexible support for structured
additive regression models, including spatial multilevel models. The models include
an IID unstructured random effect at the upper level using the `"re"` specification in
the **sx** model term (Umlauf et al. 2015); we choose the `"MCMC"` method:

```
library(R2BayesX) |> suppressPackageStartupMessages()
```

```
BX_iid <- bayesx(update(form, . ~ . + sx(NOX_ID, bs = "re")),
                family = "gaussian", data = boston_487,
                method = "MCMC", iterations = 12000,
                burnin = 2000, step = 2, seed = 123)
```

```
boston_93$BX_re <- BX_iid$effects["sx(NOX_ID):re"][[1]]$Mean
```

and the `"mrf"` (Markov Random Field) spatially structured intrinsic CAR random
effect specification based on a graph derived from converting a suitable `"nb"` object
for the upper level. The `"region id"` attribute of the `"nb"` object needs to contain
values corresponding to the indexing variable in the **sx** effects term, to facilitate the
internal construction of design matrix Z:

```
RBX_gra <- nb2gra(nb_q_93)
all.equal(row.names(RBX_gra), attr(nb_q_93, "region.id"))
# [1] TRUE
```

As we saw above in the intrinsic CAR model definition, the counts of neighbours are
entered on the diagonal, but the current implementation uses a dense, not sparse,
matrix:

```
all.equal(unname(diag(RBX_gra)), spdep::card(nb_q_93))
# [1] TRUE
```

The **sx** model term continues to include the indexing variable, and now passes through
the intrinsic CAR precision matrix:

```
BX_mrf <- bayesx(update(form, . ~ . + sx(NOX_ID, bs = "mrf",
                                         map = RBX_gra)),
                family = "gaussian", data = boston_487,
                method = "MCMC", iterations = 12000,
                burnin = 2000, step = 2, seed = 123)
```

```
boston_93$BX_ss <- BX_mrf$effects["sx(NOX_ID):mrf"][[1]]$Mean
```

16.2.4   IID, ICAR and Leroux random effects with INLA

Bivand, Gómez-Rubio, and Rue (2015) and Gómez-Rubio (2020) present the use of the **INLA** package (Rue, Lindgren, and Teixeira Krainski 2022) and the `inla` model fitting function with spatial regression models:

```
library(INLA) |> suppressPackageStartupMessages()
```

Although differing in details, the approach by updating the fixed model formula with an unstructured random effects term is very similar to that seen above:

```
INLA_iid <- inla(update(form, . ~ . + f(NOX_ID, model = "iid")),
                 family = "gaussian", data = boston_487)
```

```
boston_93$INLA_re <- INLA_iid$summary.random$NOX_ID$mean
```

As with most implementations, care is needed to match the indexing variable with the spatial weights; in this case using indices $1, \ldots, 93$ rather than the `NOX_ID` variable directly:

```
ID2 <- as.integer(as.factor(boston_487$NOX_ID))
```

The same sparse binary spatial weights matrix is used, and the intrinsic CAR representation is constructed internally:

```
INLA_ss <- inla(update(form, . ~ . + f(ID2, model = "besag",
                                        graph = W)),
                family = "gaussian", data = boston_487)
```

```
boston_93$INLA_ss <- INLA_ss$summary.random$ID2$mean
```

The sparse Leroux representation as given by Gómez-Rubio (2020) can be constructed in the following way:

```
M <- Diagonal(nrow(W), rowSums(W)) - W
Cmatrix <- Diagonal(nrow(M), 1) - M
```

This model can be estimated using the **"generic1"** model with the specified precision matrix:

```
INLA_lr <- inla(update(form, . ~ . + f(ID2, model = "generic1",
                                        Cmatrix = Cmatrix)),
                family = "gaussian", data = boston_487)
```

```
boston_93$INLA_lr <- INLA_lr$summary.random$ID2$mean
```

16.2.5 ICAR random effects with mgcv::gam()

In a very similar way, the **gam** function in the **mgcv** package (Wood 2022) can take an "mrf" term using a suitable "nb" object for the upper level. In this case the "nb" object needs to have the contents of the "region.id" attribute copied as the names of the neighbour list components, and the indexing variable needs to be a factor (Wood 2017):

```
library(mgcv)
names(nb_q_93) <- attr(nb_q_93, "region.id")
boston_487$NOX_ID <- as.factor(boston_487$NOX_ID)
```

The specification of the spatially structured term again differs in details from those above, but achieves the same purpose. The "REML" method of **bayesx** gives the same results as **gam** using "REML" in this case:

```
GAM_MRF <- gam(update(form, . ~ . + s(NOX_ID, bs = "mrf",
                                      xt = list(nb = nb_q_93))),
               data = boston_487, method = "REML")
```

The upper-level random effects may be extracted by predicting terms; as we can see, the values in all lower-level tracts belonging to the same upper-level air pollution model output zones are identical:

```
ssre <- predict(GAM_MRF, type = "terms",
                se = FALSE)[, "s(NOX_ID)"]
all(sapply(tapply(ssre, list(boston_487$NOX_ID), c),
           function(x) length(unique(round(x, 8))) == 1))
# [1] TRUE
```

so we can return the first value for each upper-level unit:

```
boston_93$GAM_ss <- aggregate(ssre, list(boston_487$NOX_ID),
                              head, n=1)$x
```

16.2.6 Upper-level random effects: summary

In the cases of **hglm**, **bayesx**, **inla** and **gam**, we could also model discrete responses without further major difficulty, and **bayesx**, **inla** and **gam** also facilitate the generalisation of functional form fitting for included covariates.

Unfortunately, the coefficient estimates for the air pollution variable for these multilevel models are not helpful. All are negative as expected, but the inclusion of the model output zone level effects, IID or spatially structured, makes it is hard to disentangle the influence of the scale of observation from that of covariates observed at that scale rather than at the tract level.

Figure 16.1 shows that the air pollution model output zone level IID random effects are very similar across the four model fitting functions reported. In all the maps, the central downtown zones have stronger negative random effect values, but strong positive values are also found in close proximity; suburban areas take values closer to zero.

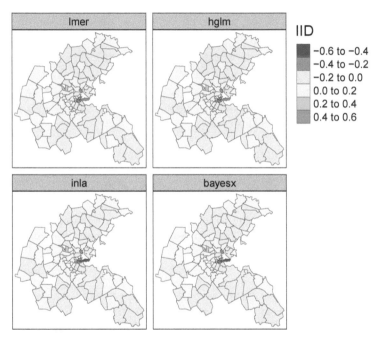

Figure 16.1: Air pollution model output zone level IID random effects estimated using **lme4**, **hglm**, **INLA** and **R2BayesX**; the range of the response, `log(median)` is 2.1893

Figure 16.2 shows that the spatially structured random effects are also very similar to each other, with the `"SAR"` spatial smooth being perhaps a little smoother than the `"CAR"` smooths when considering the range of values taken by the random effect term.

Although there is still a great need for more thorough comparative studies of model fitting functions for spatial regression including multilevel capabilities, there has been much progress over recent years. Vranckx, Neyens, and Faes (2019) offer a recent comparative survey of disease mapping spatial regression, typically set in a Poisson regression framework offset by an expected count. In Bivand and Gómez-Rubio

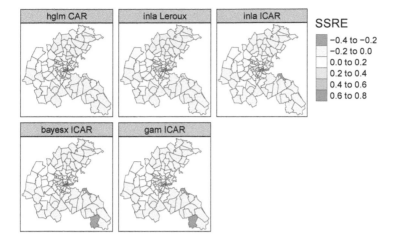

Figure 16.2: Air pollution model output zone level spatially structured random effects estimated using **hglm, HSAR, INLA, R2BayesX** and **mgcv**

(2021), methods for estimating spatial survival models using spatial weights matrices are compared with spatial probit models.

16.3 Exercises

1. Construct a multilevel dataset using the Athens housing data from the archived **HSAR** package: https://cran.r-project.org/src/contrib/Archive/HSAR/HSAR_0.5.1.tar.gz, and included in **spData** from version 2.2.1. At which point do the municipality department attribute values get copied out to all the point observations within each municipality department?

2. Create neighbour objects at both levels. Test **greensp** for spatial autocorrelation at the upper level, and then at the lower level. What has been the chief consequence of copying out the area of green spaces in square meters for the municipality departments to the point support property level?

3. Using the formula object from the vignette, assess whether adding the copied out upper-level variables seems sensible. Use mgcv::gam to fit a linear mixed effects model (IID of num_dep identifying the municipality departments) using just the lower-level variables and the lower- and upper-level variables. Do your conclusions differ?

4. Complete the analysis by replacing the IID random effects with an "mrf" Markov random field and the contiguity neighbour object created above. Do you think that it is reasonable to, for example, draw any conclusions based on the municipality department level variables such as **greensp**?

17

Spatial Econometrics Models

Spatial autoregression models using spatial weights matrices were described in some detail using maximum likelihood estimation some time ago (Cliff and Ord 1973, 1981). A family of models was elaborated in spatial econometric terms extending earlier work, and in many cases using the simultaneous autoregressive framework and row standardisation of spatial weights (Anselin 1988). The simultaneous and conditional autoregressive frameworks can be compared, and both can be supplemented using case weights to reflect the relative importance of different observations (Waller and Gotway 2004).

Before moving to presentations of issues raised in fitting spatial regression models, it is worth making a few further points. A recent review of spatial regression in a spatial econometrics setting is given by Kelejian and Piras (2017); note that their usage is to call the spatial coefficient of the lagged response λ and that of the lagged residuals ρ, the reverse of other usage (Anselin 1988; LeSage and Pace 2009); here we use ρ_{Lag} for the spatial coefficient in the spatial lag model, and ρ_{Err} for the spatial error model. One interesting finding is that relatively dense spatial weights matrices may down-weight model estimates, suggesting that sparser weights are preferable (Smith 2009). Another useful finding is that the presence of residual spatial autocorrelation need not bias the estimates of variance of regression coefficients, provided that the covariates themselves do not exhibit spatial autocorrelation (Smith and Lee 2012). In general, however, the footprints of the spatial processes of the response and covariates may not be aligned, and if covariates and the residual are autocorrelated, it is likely that the estimates of variance of regression coefficients will be biased downwards if attempts are not made to model the spatial processes.

17.1 Spatial econometric models: definitions

In trying to model spatial processes, one of the earliest spatial econometric representations is to model the spatial autocorrelation in the residual (spatial error model, SEM):

$$\mathbf{y} = \mathbf{X}\beta + \mathbf{u}, \qquad \mathbf{u} = \rho_{\text{Err}}\mathbf{W}\mathbf{u} + \varepsilon,$$

where $\mathbf{y}$ is an $(N \times 1)$ vector of observations on a response variable taken at each of N locations, $\mathbf{X}$ is an $(N \times k)$ matrix of covariates, β is a $(k \times 1)$ vector of parameters, $\mathbf{u}$ is an $(N \times 1)$ spatially autocorrelated disturbance vector, ε is an $(N \times 1)$ vector of independent and identically distributed disturbances and ρ_{Err} is a scalar spatial parameter.

This model, and other spatial econometric models, do not fit into the mixed models framework. Here the modelled spatial process interacts directly with the response, covariates, and their coefficients. This modelling framework appears to draw on an older tradition extending time series to two dimensions:

$$\mathbf{u} = (\mathbf{I} - \rho_{\mathrm{Err}}\mathbf{W})^{-1}\varepsilon, \quad \mathbf{y} = \mathbf{X}\beta + (\mathbf{I} - \rho_{\mathrm{Err}}\mathbf{W})^{-1}\varepsilon, \quad (\mathbf{I} - \rho_{\mathrm{Err}}\mathbf{W})\mathbf{y} = (\mathbf{I} - \rho_{\mathrm{Err}}\mathbf{W})\mathbf{X}\beta + \varepsilon.$$

If the processes in the covariates and the response match, we should find little difference between the coefficients of a least squares and a SEM, but very often they diverge, suggesting that a Hausman test for this condition should be employed (Pace and LeSage 2008). This may be related to earlier discussions of a spatial equivalent to the unit root and cointegration where spatial processes match (Fingleton 1999).

A model with a spatial process in the response only is termed a spatial lag model (SLM, often SAR - spatial autoregressive) (LeSage and Pace 2009). Durbin models add the spatially lagged covariates to the covariates included in the spatial model; spatial Durbin models are reviewed by Mur and Angulo (2006). If it is chosen to admit a spatial process in the residuals in addition to a spatial process in the response, again two models are formed, a general nested model (GNM) nesting all the others, and a model without spatially lagged covariates (SAC, also known as SARAR - Spatial AutoRegressive-AutoRegressive model). If neither the residuals nor the response are modelled with spatial processes, spatially lagged covariates may be added to a linear model, as a spatially lagged X model (SLX) (Elhorst 2010; Bivand 2012; LeSage 2014; Halleck Vega and Elhorst 2015). We can write the GNM as:

$$\mathbf{y} = \rho_{\mathrm{Lag}}\mathbf{W}\mathbf{y} + \mathbf{X}\beta + \mathbf{W}\mathbf{X}\gamma + \mathbf{u}, \qquad \mathbf{u} = \rho_{\mathrm{Err}}\mathbf{W}\mathbf{u} + \varepsilon,$$

where γ is a $(k' \times 1)$ vector of parameters. k' defines the subset of the intercept and covariates, often $k' = k - 1$ when using row standardised spatial weights and omitting the spatially lagged intercept.

This may be constrained to the double spatial coefficient model SAC/SARAR by setting $\gamma = 0$, to the spatial Durbin (SDM) by setting $\rho_{\mathrm{Err}} = 0$, and to the error Durbin model (SDEM) by setting $\rho_{\mathrm{Lag}} = 0$. Imposing more conditions gives the spatial lag model (SLM) with $\gamma = 0$ and $\rho_{\mathrm{Err}} = 0$, the SEM with $\gamma = 0$ and $\rho_{\mathrm{Lag}} = 0$, and the SLX with $\rho_{\mathrm{Lag}} = 0$ and $\rho_{\mathrm{Err}} = 0$.

Although making predictions for new locations for which covariates are observed was raised as an issue some time ago, it has taken many years to make progress in reviewing the possibilities (Bivand 2002; Goulard, Laurent, and Thomas-Agnan 2017; Laurent and Margaretic 2021). The prediction methods for SLM, SDM, SEM, SDEM, SAC, and GNM models fitted with maximum likelihood were contributed as a Google Summer of Coding project by Martin Gubri. This work, and work on similar models with missing data (Suesse 2018) is also relevant for exploring censored median house values in the Boston dataset. Work on prediction also exposed the importance of the reduced form of these models, in which the spatial process in the response interacts with the regression coefficients in the SLM, SDM, SAC and GNM models.

The consequence of these interactions is that a unit change in a covariate will only impact the response as the value of the regression coefficient if the spatial coefficient of the lagged response is zero. Where it is non-zero, global spillovers, impacts, come into play, and these impacts should be reported rather than the regression coefficients (LeSage and Pace 2009; Elhorst 2010; Bivand 2012; LeSage 2014; Halleck Vega and Elhorst 2015). Local impacts may be reported for SDEM and SLX models, using a linear combination to calculate standard errors for the total impacts of each covariate (sums of coefficients on the covariates and their spatial lags).

This can be seen from the GNM data generation process:

$$(\mathbf{I} - \rho_{\mathrm{Err}}\mathbf{W})(\mathbf{I} - \rho_{\mathrm{Lag}}\mathbf{W})\mathbf{y} = (\mathbf{I} - \rho_{\mathrm{Err}}\mathbf{W})(\mathbf{X}\beta + \mathbf{W}\mathbf{X}\gamma) + \varepsilon,$$

re-writing:

$$\mathbf{y} = (\mathbf{I} - \rho_{\mathrm{Lag}}\mathbf{W})^{-1}(\mathbf{X}\beta + \mathbf{W}\mathbf{X}\gamma) + (\mathbf{I} - \rho_{\mathrm{Lag}}\mathbf{W})^{-1}(\mathbf{I} - \rho_{\mathrm{Err}}\mathbf{W})^{-1}\varepsilon.$$

There is interaction between the ρ_{Lag} and β (and γ if present) coefficients. This can be seen from the partial derivatives: $\partial y_i / \partial x_{jr} = ((\mathbf{I} - \rho_{\mathrm{Lag}}\mathbf{W})^{-1}(\mathbf{I}\beta_r + \mathbf{W}\gamma_r))_{ij}$. This dense matrix $S_r(\mathbf{W}) = ((\mathbf{I} - \rho_{\mathrm{Lag}}\mathbf{W})^{-1}(\mathbf{I}\beta_r + \mathbf{W}\gamma_r))$ expresses the direct impacts (effects) on its principal diagonal, and indirect impacts in off-diagonal elements.

Piras and Prucha (2014) revisit and correct Raymond J. G. M. Florax, Folmer, and Rey (2003) (see also comments by Hendry (2006) and Raymond J. G. M. Florax, Folmer, and Rey (2006)), finding that the common use of pre-test strategies for model selection probably ought to be replaced by the estimation of the most general model appropriate for the relationships being modelled. In light of this finding, pre-test model selection will not be used here.

Current work in the **spatialreg** package is focused on refining the handling of spatially lagged covariates using a consistent `Durbin=` argument taking either a logical value or a formula giving the subset of covariates to add in spatially lagged form. There is a speculation that some covariates, for example some dummy variables, should not be added in spatially lagged form. This then extends to handling these included spatially lagged covariates appropriately in calculating impacts. This work applies to cross-sectional models fitted using MCMC or maximum likelihood, and will offer facilities to spatial panel models.

It is worth mentioning the almost unexplored issues of functional form assumptions, for which flexible structures are useful, including spatial quantile regression presented in the **McSpatial** package (McMillen 2013). There are further issues with discrete response variables, covered by some functions in **McSpatial**, by the new package **spldv** (Sarrias and Piras 2022), and in the **spatialprobit** and **ProbitSpatial** packages (Wilhelm and Matos 2013; Martinetti and Geniaux 2017); the MCMC implementations of the former are based on LeSage and Pace (2009). Finally, Wagner and Zeileis (2019) show how an SLM model may be used in the setting of recursive partitioning, with an implementation using `spatialreg::lagsarlm()` in the **lagsarlmtree** package.

The review of cross-sectional maximum likelihood and generalised method of moments (GMM) estimators in **spatialreg** (Bivand and Piras 2022) and **sphet** for spatial econometrics style spatial regression models by Bivand and Piras (2015) is still largely valid. In the review, estimators in these R packages were compared with alternative implementations available in other programming languages elsewhere. The review did not cover Bayesian spatial econometrics style spatial regression. More has changed with respect to spatial panel estimators described in Millo and Piras (2012) but will not be covered here.

Because Bivand, Millo, and Piras (2021) covers many of the features of R packages for spatial econometrics, updating Bivand and Piras (2015), and including recent advances in General Method of Moments and spatial panel modelling, this chapter will be restricted to a small number of examples drawing on Bivand (2017) using the Boston house value dataset.

17.2   Maximum likelihood estimation in spatialreg

For models with single spatial coefficients (SEM and SDEM using `errorsarlm()`, SLM and SDM using `lagsarlm()`), the methods initially described by Ord (1975) are used. The following table shows the functions that can be used to estimate the models described above using maximum likelihood.

model	model name	maximum likelihood estimation function
SEM	spatial error	`errorsarlm(..., Durbin=FALSE)`
SEM	spatial error	`spautolm(..., family="SAR")`
SDEM	spatial Durbin error	`errorsarlm(..., Durbin=TRUE)`
SLM	spatial lag	`lagsarlm(..., Durbin=FALSE)`
SDM	spatial Durbin	`lagsarlm(..., Durbin=TRUE)`
SAC	spatial autoregressive combined	`sacsarlm(..., Durbin=FALSE)`
GNM	general nested	`sacsarlm(..., Durbin=TRUE)`

The estimating functions `errorsarlm()` and `lagsarlm()` take similar arguments, where the first two, `formula=` and `data=` are shared by most model estimating functions. The third argument is a `listw` spatial weights object, while `na.action=` behaves as in other model estimating functions if the spatial weights can reasonably be subsetted to avoid observations with missing values. The `weights=` argument may be used to provide weights indicating the known degree of per-observation variability in the variance term - this is not available for `lagsarlm()`.

The `Durbin=` argument replaces the earlier `type=` and `etype=` arguments, and if not given is taken as `FALSE`. If given, it may be `FALSE`, `TRUE` in which case all spatially lagged covariates are included, or a one-sided formula specifying which spatially lagged covariates should be included. The `method=` argument gives the method for calculating the log determinant term in the log likelihood function, and defaults to `"eigen"`, suitable for moderately sized datasets. The `interval=` argument gives the bounds of the domain for the line search using `stats::optimize()` used for finding the spatial coefficient. The `tol.solve()` argument, passed through to `base::solve()`, was needed to handle datasets with differing numerical scales among the coefficients which hindered inversion of the variance-covariance matrix; the default value in `base::solve()` used to be much larger. The `control=` argument takes a list of control values to permit more careful adjustment of the running of the estimation function.

The `sacsarlm()` function may take second spatial weights and interval arguments if the spatial weights used to model the two spatial processes in the SAC and GNM specifications differ. By default, the same spatial weights are used. By default, `stats::nlminb()` is used for numerical optimisation, using a heuristic to choose starting values. Like `lagsarlm()`, this function does not take a `weights=` argument.

Where larger datasets are used, a numerical Hessian approach is used to calculate the variance-covariance matrix of coefficients, rather than an analytical asymptotic approach.

17.2.1 Boston house value dataset examples

The examples use the objects read and created in Chapter 16, based on Bivand (2017).

```
library(spatialreg)
eigs_489 <- eigenw(lw_q_489)
SDEM_489 <- errorsarlm(form, data = boston_489,
        listw = lw_q_489, Durbin = TRUE, zero.policy = TRUE,
        control = list(pre_eig = eigs_489))
SEM_489 <- errorsarlm(form, data = boston_489,
        listw = lw_q_489, zero.policy = TRUE,
        control = list(pre_eig = eigs_489))
```

Here we are using the `control=` list argument to pass through pre-computed eigenvalues for the default **"eigen"** method.

```
cbind(data.frame(model=c("SEM", "SDEM")),
        rbind(broom::tidy(Hausman.test(SEM_489)),
            broom::tidy(Hausman.test(SDEM_489))))[,1:4]
#    model statistic  p.value parameter
# 1    SEM      52.0 2.83e-06        14
# 2   SDEM      48.7 6.48e-03        27
```

Both Hausman test results for the 489 tract dataset suggest that the regression coefficients do differ from their non-spatial counterparts, perhaps indicating that the footprints of the spatial processes do not match.

```
eigs_94 <- eigenw(lw_q_94)
SDEM_94 <- errorsarlm(form, data=boston_94, listw=lw_q_94,
                    Durbin = TRUE,
                    control = list(pre_eig=eigs_94))
SEM_94 <- errorsarlm(form, data = boston_94, listw = lw_q_94,
                    control = list(pre_eig = eigs_94))
```

For the 94 air pollution model output zones, the Hausman tests find little difference between coefficients:

```
cbind(data.frame(model=c("SEM", "SDEM")),
      rbind(broom::tidy(Hausman.test(SEM_94)),
            broom::tidy(Hausman.test(SDEM_94))))[, 1:4]
#    model statistic p.value parameter
# 1    SEM     15.66   0.335        14
# 2   SDEM      9.21   0.999        27
```

This is related to the fact that the SEM and SDEM models add little to least squares or SLX at the air pollution model output zone level, using likelihood ratio tests:

```
cbind(data.frame(model=c("SEM", "SDEM")),
      rbind(broom::tidy(LR1.Sarlm(SEM_94)),
            broom::tidy(LR1.Sarlm(SDEM_94))))[,c(1, 4:6)]
#    model statistic p.value parameter
# 1    SEM     2.593   0.107         1
# 2   SDEM     0.216   0.642         1
```

We can use **spatialreg::LR.Sarlm()** to apply a likelihood ratio test between nested models, but here choose **lmtest::lrtest()**, which gives the same results, preferring models including spatially lagged covariates both for tracts and model output zones:

```
o <- lmtest::lrtest(SEM_489, SDEM_489)
attr(o, "heading")[2] <- "Model 1: SEM_489\nModel 2: SDEM_489"
o
# Likelihood ratio test
#
# Model 1: SEM_489
# Model 2: SDEM_489
#   #Df LogLik Df Chisq Pr(>Chisq)
# 1  16    274
# 2  29    311 13  74.4    1.2e-10 ***
# ---
# Signif. codes:  0 '***' 0.001 '**' 0.01 '*' 0.05 '.' 0.1 ' ' 1


o <- lmtest::lrtest(SEM_94, SDEM_94)
attr(o, "heading")[2] <- "Model 1: SEM_94\nModel 2: SDEM_94"
o
# Likelihood ratio test
#
# Model 1: SEM_94
# Model 2: SDEM_94
#   #Df LogLik Df Chisq Pr(>Chisq)
# 1  16   59.7
```

```
# 2  29   81.3 13  43.2     4.2e-05 ***
# ---
# Signif. codes:  0 '***' 0.001 '**' 0.01 '*' 0.05 '.' 0.1 ' ' 1
```

The SLX model is fitted using least squares and also returns a log likelihood value, letting us test whether we need a spatial process in the residuals. In the tract dataset, we obviously do:

```
SLX_489 <- lmSLX(form, data = boston_489, listw = lw_q_489,
                  zero.policy = TRUE)
o <- lmtest::lrtest(SLX_489, SDEM_489)
attr(o, "heading")[2] <- "Model 1: SLX_489\nModel 2: SDEM_489"
o
# Likelihood ratio test
#
# Model 1: SLX_489
# Model 2: SDEM_489
#    #Df LogLik Df Chisq Pr(>Chisq)
# 1  28    231
# 2  29    311  1   159     <2e-16 ***
# ---
# Signif. codes:  0 '***' 0.001 '**' 0.01 '*' 0.05 '.' 0.1 ' ' 1
```

but in the output zone case, we do not.

```
SLX_94 <- lmSLX(form, data = boston_94, listw = lw_q_94)
o <- lmtest::lrtest(SLX_94, SDEM_94)
attr(o, "heading")[2] <- "Model 1: SLX_94\nModel 2: SDEM_94"
o
# Likelihood ratio test
#
# Model 1: SLX_94
# Model 2: SDEM_94
#    #Df LogLik Df Chisq Pr(>Chisq)
# 1  28    81.2
# 2  29    81.3  1  0.22       0.64
```

These outcomes are sustained also when we use the counts of house units by tract and output zones as case weights:

```
SLX_489w <- lmSLX(form, data = boston_489, listw = lw_q_489,
                  weights = units, zero.policy = TRUE)
SDEM_489w <- errorsarlm(form, data = boston_489,
                        listw = lw_q_489, Durbin = TRUE,
                        weights = units, zero.policy = TRUE,
                        control = list(pre_eig = eigs_489))
```

```
o <- lmtest::lrtest(SLX_489w, SDEM_489w)
attr(o, "heading")[2] <- "Model 1: SLX_489w\nModel 2: SDEM_489w"
o
# Likelihood ratio test
#
# Model 1: SLX_489w
# Model 2: SDEM_489w
#   #Df LogLik Df Chisq Pr(>Chisq)
# 1  28    311
# 2  29    379  1   136    <2e-16 ***
# ---
# Signif. codes:  0 '***' 0.001 '**' 0.01 '*' 0.05 '.' 0.1 ' ' 1


SLX_94w <- lmSLX(form, data = boston_94, listw = lw_q_94,
                 weights = units)
SDEM_94w <- errorsarlm(form, data = boston_94, listw = lw_q_94,
                       Durbin = TRUE, weights = units,
                       control = list(pre_eig = eigs_94))
o <- lmtest::lrtest(SLX_94w, SDEM_94w)
attr(o, "heading")[2] <- "Model 1: SLX_94w\nModel 2: SDEM_94w"
o
# Likelihood ratio test
#
# Model 1: SLX_94w
# Model 2: SDEM_94w
#   #Df LogLik Df Chisq Pr(>Chisq)
# 1  28   97.5
# 2  29   98.0  1  0.92       0.34
```

In this case and based on arguments advanced in Bivand (2017), the use of weights is justified because tract counts of reported housing units underlying the weighted median values vary from 5 to 3,031, and air pollution model output zone counts vary from 25 to 12,411. Because of this, and because a weighted general nested model has not been developed, we cannot take the GNM as the starting point for general-to-simpler testing, but we start rather from the SDEM model and use the Hausman test to guide the choice of units of observation.

17.3 Impacts

Global impacts have been seen as crucial for reporting results from fitting models including the spatially lagged response (SLM, SDM, SAC, GNM) for over 10 years (LeSage and Pace 2009). Extension to other models including spatially lagged covariates (SLX, SDEM) has followed (Elhorst 2010; Bivand 2012; Halleck Vega and Elhorst 2015). For SLM, SDM, SAC, and GNM models fitted with maximum likelihood or

GMM, the variance-covariance matrix of the coefficients is available, and can be used to make random draws from a multivariate Normal distribution with mean set to coefficient values and variance to the estimated variance-covariance matrix. For these models fitted using Bayesian methods, draws are already available. In the SDEM case, the draws on the regression coefficients of the unlagged covariates represent direct impacts, and draws on the coefficients of the spatially lagged covariates represent indirect impacts, and their by-draw sums the total impacts.

Since sampling is not required for inference for SLX and SDEM models, linear combination is used for models fitted using maximum likelihood; results are shown here for the air pollution variable only. The literature has not yet resolved the question of how to report model output, as each covariate is now represented by three impacts. Where spatially lagged covariates are included, two coefficients are replaced by three impacts, here for the air pollution variable of interest.

```
sum_imp_94_SDEM <- summary(impacts(SDEM_94))
rbind(Impacts = sum_imp_94_SDEM$mat[5,],
      SE = sum_imp_94_SDEM$semat[5,])
#            Direct Indirect    Total
# Impacts -0.01276 -0.01845 -0.0312
# SE       0.00235  0.00472  0.0053
```

In the SLX and SDEM models, the direct impacts are the consequences for the response of changes in air pollution in the same observational entity, and the indirect (local) impacts are the consequences for the response of changes in air pollution in neighbouring observational entities.

```
sum_imp_94_SLX <- summary(impacts(SLX_94))
rbind(Impacts = sum_imp_94_SLX$mat[5,],
      SE = sum_imp_94_SLX$semat[5,])
#            Direct Indirect    Total
# Impacts -0.0128 -0.01874 -0.03151
# SE       0.0028  0.00556  0.00611
```

Applying the same approaches to the weighted spatial regressions, the total impacts of air pollution on house values are reduced, but remain significant:

```
sum_imp_94_SDEMw <- summary(impacts(SDEM_94w))
rbind(Impacts = sum_imp_94_SDEMw$mat[5,],
      SE = sum_imp_94_SDEMw$semat[5,])
#            Direct Indirect    Total
# Impacts -0.00592 -0.01076 -0.01668
# SE       0.00269  0.00531  0.00559
```

On balance, using a weighted spatial regression representation including only the spatially lagged covariates aggregated to the air pollution model output zone level seems to clear most of the misspecification issues, and as Bivand (2017) discusses in more detail, it gives a willingness to pay for pollution abatement that is much larger than misspecified alternative models:

```
sum_imp_94_SLXw <- summary(impacts(SLX_94w))
rbind(Impacts = sum_imp_94_SLXw$mat[5,],
      SE = sum_imp_94_SLXw$semat[5,])
#              Direct Indirect    Total
# Impacts -0.00620 -0.01221 -0.01842
# SE       0.00326  0.00628  0.00629
```

17.4 Predictions

In the Boston tracts dataset, 17 observations of median house values, the response, are censored. We will use the **predict()** method for **"Sarlm"** objects to fill in these values; the method was rewritten by Martin Gubri based on Goulard, Laurent, and Thomas-Agnan (2017), see also Laurent and Margaretic (2021). The **pred.type=** argument specifies the prediction strategy among those presented in the article.

Using these as an example and comparing some **pred.type=** variants for the SDEM model and predicting out-of-sample, we can see that there are differences, suggesting that this is a fruitful area for study. There have been a number of alternative proposals for handling missing variables (Gómez-Rubio, Bivand, and Rue 2015; Suesse 2018). Another reason for increasing attention on prediction is that it is fundamental for machine learning approaches, in which prediction for validation and test datasets drives model specification choice. The choice of training and other datasets with dependent spatial data remains an open question, and, is certainly not as simple as with independent data.

Here, we'll list the predictions for the censored tract observations using three different prediction types, taking the exponent to get back to the USD median house values. Note that the **row.names()** of the **newdata=** object are matched with the whole-data spatial weights matrix **"region.id"** attribute to make out-of-sample prediction possible:

```
nd <- boston_506[is.na(boston_506$median),]
t0 <- exp(predict(SDEM_489, newdata = nd, listw = lw_q,
                  pred.type = "TS", zero.policy =TRUE))
suppressWarnings(t1 <- exp(predict(SDEM_489, newdata = nd,
                           listw = lw_q,
                           pred.type = "KP2",
                           zero.policy = TRUE)))
suppressWarnings(t2 <- exp(predict(SDEM_489, newdata = nd,
                           listw = lw_q,
```

```
                                    pred.type = "KP5",
                                    zero.policy = TRUE)))
```

We can also use the **"slm"** model in INLA to predict missing response values as part
of the model fitting function call. A certain amount of set-up code is required as the
"slm" model is still experimental:

```
library(INLA)
# Loading required package: foreach
# Loading required package: parallel
# Loading required package: sp
# This is INLA_22.05.07 built 2022-05-07 09:52:03
#                    UTC.
#  - See www.r-inla.org/contact-us for how to get help.
#  - To enable PARDISO sparse library; see inla.pardiso()
#
# Attaching package: 'INLA'
# The following object is masked _by_ '.GlobalEnv':
#
#     f
W <- as(lw_q, "CsparseMatrix")
n <- nrow(W)
e <- eigenw(lw_q)
re.idx <- which(abs(Im(e)) < 1e-6)
rho.max <- 1 / max(Re(e[re.idx]))
rho.min <- 1 / min(Re(e[re.idx]))
rho <- mean(c(rho.min, rho.max))
boston_506$idx <- 1:n
zero.variance = list(prec = list(initial = 25, fixed = TRUE))
args.slm <- list(rho.min = rho.min, rho.max = rho.max, W = W,
                 X = matrix(0, n, 0), Q.beta = matrix(1,0,0))
hyper.slm <- list(prec = list(prior = "loggamma",
                              param = c(0.01, 0.01)),
                  rho = list(initial = 0, prior = "logitbeta",
                             param = c(1,1)))
WX <- create_WX(model.matrix(update(form, CMEDV ~ .),
                             data = boston_506), lw_q)
SDEM_506_slm <- inla(update(form,
                            . ~ . + WX + f(idx, model = "slm",
                                           args.slm = args.slm,
                                           hyper = hyper.slm)),
                data = boston_506, family = "gaussian",
                control.family = list(hyper = zero.variance),
                control.compute = list(dic = TRUE, cpo = TRUE))
mv_mean <- exp(SDEM_506_slm$summary.fitted.values$mean[
               which(is.na(boston_506$median))])
```

INLA also provide gridded estimates of the marginal distributions of the predictions, offering a way to assess the uncertainty associated with the predicted values.

```
data.frame(fit_TS = t0[,1], fit_KP2 = c(t1), fit_KP5 = c(t2),
    INLA_slm = mv_mean, censored =
    boston_506$censored[as.integer(attr(t0, "region.id"))])
#      fit_TS fit_KP2 fit_KP5 INLA_slm censored
# 13    23912   29477   28147    31094    right
# 14    28126   27001   28516    31376    right
# 15    30553   36184   32476    41034    right
# 17    18518   19621   18878    21077    right
# 43     9564    6817    7561     6875     left
# 50     8371    7196    7383     6891     left
# 312   51477   53301   54173    56314    right
# 313   45921   45823   47095    46526    right
# 314   44196   44586   45361    42839    right
# 317   43427   45707   45442    47969    right
# 337   39879   42072   41127    41422    right
# 346   44708   46694   46108    45805    right
# 355   48188   49068   48911    49116    right
# 376   42881   45883   44966    47657    right
# 408   44294   44615   45670    46281    right
# 418   38211   43375   41914    43799    right
# 434   41647   41690   42398    41526    right
```

The spatial regression toolbox remains incomplete, and it will take time to fill in blanks. It remains unfortunate that the several traditions in spatial regression seldom seem to draw on each others' understandings and advances.

17.5 Exercises

1. Referring to Piras and Prucha (2014) and Raymond J. G. M. Florax, Folmer, and Rey (2003), if we choose to use a pre-test strategy, do linear models of the properties-only dataset and the properties with added municipality department variables show residual spatial dependence? Which model specifications might the pre-tests indicate?

2. Could the inclusion of municipality department dummies, or a municipality department regimes model assist in reducing residual spatial dependence?

3. Attempt to fit a SEM specification by maximum likelihood (see Bivand, Millo, and Piras (2021) for GMM code examples) to the properties-only and the properties with added municipality department variables models; extend to an SDEM model. Repeat with SLX models; how might the changes in the tests of residual autocorrelation in the SLX models be interpreted? How might you interpret the highly significant outcomes of Hausman tests on the SEM and SDEM models?

4. Fit GNM specifications to the properties-only and the properties with added municipality department variables models; can these models be simplified to say SDM or SDEM representations?
5. Do the model estimates reached in the Chapter 16 exercises provide more clarity than those in this chapter?

A

Older R Spatial Packages

A.1 Retiring rgdal and rgeos

R users who have been around a bit longer, in particular before packages like **sf** and **stars** were developed, may be more familiar with older packages like **maptools**, **sp**, **rgeos**, and **rgdal**. A fair question is whether they should migrate existing code and/or existing R packages depending on these packages. The answer is: yes.

Packages **maptools**, **rgdal**, and **rgeos** will retire during 2023. Retirement means that maintenance will halt, and that as a consequence the packages will be archived on CRAN. The source code repositories on R-Forge will remain as long as R-Forge does itself. One reason for retirement is that their maintainer has retired, a more important reason that their role has been superseded by the newer packages. We hold it most unlikely that a new maintainer will take over the R-Forge repositories, in part because much of the code of these packages has gradually evolved along with developments in the GEOS, GDAL, and PROJ libraries, and contains numerous constructs that are outdated and make it forbidding to read.

Before **rgeos** and **rgdal** retire, existing ties that package **sp** has to **rgdal** and **rgeos** can and will be replaced by ties to package **sf**. This involves for example validation of coordinate reference system identifiers, and checking whether rings are holes or exterior rings. Chosen **maptools** functions may also be moved to **sp**.

A.2 Links and differences between sf and sp

There are a number of differences between **sf** and **sp**. The most notable is that **sp** classes are formal, S4 classes where **sf** uses the (more) informal S3 class hierarchy. **sf** objects are derived from data.frames or tibbles and because of that are more readily interfaceable with much of the existing R ecosystem, especially with the tidyverse package family. **sf** objects keep geometry in a list-column, meaning that a geometry is *always* a list element. Package **sp** used data structures much less strictly, and for instance all coordinates of `SpatialPoints` or `SpatialPixels` are kept in matrices, which is much more performant for certain problems but is not possible with a list-column. Conversion from an **sf** object **x** to its **sp** equivalent is done by

```
library(sp)
y = as(x, "Spatial")
```

and the conversion the other way around is done by

```
x0 = st_as_sf(y)
```

There are some limitations to conversions like this:

- **sp** does not distinguish between `LINESTRING` and `MULTILINESTRING` geometries, or between `POLYGON` or `MULTIPOLYGON`. For example, a `LINESTRING` will after conversion to **sp** come back as a `MULTILINESTRING`
- **sp** does not have a representation for `GEOMETRYCOLLECTION` geometries, or **sf** objects with geometry types *not* in the "big seven" (Section 3.1.1)
- **sf** or **sfc** objects of geometry type `GEOMETRY`, with mixed geometry types, cannot be converted into **sp** objects
- attribute-geometry relationship attributes get lost when converting to **sp**
- **sf** objects with more than one geometry list-column will, when converting to **sp**, lose their secondary list-column(s).

A.3   Migration code and packages

The wiki page of the GitHub site for **sf**, found at https://github.com/r-spatial/sf/wiki/Migrating contains a list of methods and functions in **rgeos**, **rgdal**, and **sp** and the corresponding **sf** method or function. This may help converting existing code or packages.

A simple approach to migrate code is when only `rgdal::readOGR` is used to read `file`. As an alternative, one might use

```
x = as(sf::read_sf("file"), "Spatial")
```

however possible arguments to `readOGR`, when used, would need more care. An effort by us is underway to convert all code of our earlier book *Applied Spatial Data Analysis with R* (with Virgilio Gómez-Rubio, Bivand, Pebesma, and Gómez-Rubio (2013)) to run entirely without **rgdal**, **rgeos**, and **maptools** and where possible without **sp**. The scripts are found at https://github.com/rsbivand/sf_asdar2ed.

A.4   Package raster and terra

Package **raster** has been a workhorse package for analysing raster data with R since 2010, and has since grown into a package for "Geographic Data Analysis and

Modeling" (Hijmans 2023a), indicating that it is used for all kinds of spatial data. The **raster** package uses **sp** objects for vector data, and **terra** to read and write data to formats served by the GDAL library. Its successor package **terra**, for "Spatial Data Analysis" (Hijmans 2023b), "is very similar to the **raster** package; but [...] can do more, is easier to use, and [...] is faster". The **terra** package comes with its own classes for vector data, but accepts many **sf** objects, with similar restrictions as listed above for conversion to **sp**. Package **terra** has its own direct links to GDAL, GEOS, and PROJ, so, no longer needs other packages for that.

Raster maps, or stacks of them from package **raster** or **terra** can be converted to **stars** objects using `st_as_stars()`. Package **sf** contains an `st_as_sf()` method for `SpatVector` objects from package **terra**.

The online book *Spatial Data Science with R*, written by Robert Hijmans and found at https://rspatial.org/terra details the **terra** approach to spatial data analysis. Package **sf** and **stars** and several other r-spatial packages discussed in this book reside on the **r-spatial** GitHub organisation (note the hyphen between **r** and **spatial**, which is absent on Hijmans' organisation), which has a blog site, with links to this book, found at https://r-spatial.org/book.

Packages **sf** and **stars** on one hand and **terra** on the other have many goals in common, but try to reach them in slightly different ways, emphasising different aspects of data analysis, software engineering, and community management. Although this may confuse some users, we believe that these differences enrich the R package ecosystem, are beneficial to users, encourage diversity and choice, and hopefully work as an encouragement for others to continue trying out new ideas when using R for spatial data problems, and to help carrying the R spatial flag.

B

R Basics

This chapter provides some minimal R basics that may make it easier to read this book. A more comprehensive book on R basics is given in Wickham (2014a), chapter 2.

B.1 Pipes

The |> (pipe) symbols should be read as *then:* we read

```
a |> b() |> c() |> d(n = 10)
```

as *with a do b, then c, then d with n being 10*, and that is just alternative syntax for

```
d(c(b(a)), n = 10)
```

or

```
tmp1 <- b(a)
tmp2 <- c(tmp1)
tmp3 <- d(tmp2, n = 10)
```

To many, the pipe-form is easier to read because execution order follows reading order, from left to right. Like nested function calls, it avoids the need to choose names for intermediate results. As with nested function calls, it is hard to debug intermediate results that diverge from our expectations. Note that the intermediate results do exist in memory, so neither form saves memory allocation. The |> native pipe that appeared in R 4.1.0 as used in this book, can be safely substituted by the %>% pipe of package **magrittr**.

B.2   Data structures

As pointed out by Chambers (2016), *everything that exists in R is an object.* This includes objects that make things happen, such as language objects or functions, but also the more basic "things", such as data objects. Some basic R data structures will now be discussed.

B.2.1   Homogeneous vectors

Data objects contain data, and possibly metadata. Data is always in the form of a vector, which can have different types. We can find the type by **typeof**, and vector length by **length**. Vectors are created by **c**, which combines individual elements:

```
typeof(1:10)
# [1] "integer"
length(1:10)
# [1] 10
typeof(1.0)
# [1] "double"
length(1.0)
# [1] 1
typeof(c("foo", "bar"))
# [1] "character"
length(c("foo", "bar"))
# [1] 2
typeof(c(TRUE, FALSE))
# [1] "logical"
```

Vectors of this kind can only have a single type.

Note that vectors can have zero length:

```
i <- integer(0)
typeof(i)
# [1] "integer"
i
# integer(0)
length(i)
# [1] 0
```

We can retrieve (or in assignments: replace) elements in a vector using [or [[:

```
a <- c(1,2,3)
a[2]
# [1] 2
a[[2]]
# [1] 2
a[2:3]
# [1] 2 3
a[2:3] <- c(5,6)
a
# [1] 1 5 6
a[[3]] <- 10
a
# [1]  1  5 10
```

where the difference is that [can operate on an index *range* (or multiple indexes), and [[operates on a single vector value.

B.2.2 Heterogeneous vectors: list

An additional vector type is the list, which can combine any types in its elements:

```
l <- list(3, TRUE, "foo")
typeof(l)
# [1] "list"
length(l)
# [1] 3
```

For lists, there is a further distinction between [and [[: the single [returns always a list, and [[returns the *contents* of a list element:

```
l[1]
# [[1]]
# [1] 3
l[[1]]
# [1] 3
```

For replacement, one case use [when providing a list, and [[when providing a new value:

```
l[1:2] <- list(4, FALSE)
l
# [[1]]
# [1] 4
#
```

```
# [[2]]
# [1] FALSE
#
# [[3]]
# [1] "foo"
l[[3]] <- "bar"
l
# [[1]]
# [1] 4
#
# [[2]]
# [1] FALSE
#
# [[3]]
# [1] "bar"
```

In case list elements are *named*, as in

```
l <- list(first = 3, second = TRUE, third = "foo")
l
# $first
# [1] 3
#
# $second
# [1] TRUE
#
# $third
# [1] "foo"
```

we can use names as in `l[["second"]]` and this can be abbreviated to

```
l$second
# [1] TRUE
l$second <- FALSE
l
# $first
# [1] 3
#
# $second
# [1] FALSE
#
# $third
# [1] "foo"
```

This is convenient, but it also requires name look-up in the names attribute (see below).

B.2.3 NULL and removing list elements

NULL is the null value in R; it is special in the sense that it doesn't work in simple comparisons:

```
3 == NULL # not FALSE!
# logical(0)
NULL == NULL # not even TRUE!
# logical(0)
```

but has to be treated specially, using **is.null**:

```
is.null(NULL)
# [1] TRUE
```

When we want to remove one or more list elements, we can do so by creating a new list that does not contain the elements that needed removal, as in

```
l <- l[c(1,3)] # remove second, implicitly
l
# $first
# [1] 3
#
# $third
# [1] "foo"
```

but we can also assign NULL to the element we want to eliminate:

```
l$second <- NULL
l
# $first
# [1] 3
#
# $third
# [1] "foo"
```

B.2.4 Attributes

We can glue arbitrary metadata objects to data objects, as in

```
a <- 1:3
attr(a, "some_meta_data") = "foo"
a
# [1] 1 2 3
# attr(,"some_meta_data")
# [1] "foo"
```

and this can be retrieved, or replaced by

```
attr(a, "some_meta_data")
# [1] "foo"
attr(a, "some_meta_data") <- "bar"
attr(a, "some_meta_data")
# [1] "bar"
```

In essence, the attribute of an object is a named list, and we can get or set the complete list by

```
attributes(a)
# $some_meta_data
# [1] "bar"
attributes(a) = list(some_meta_data = "foo")
attributes(a)
# $some_meta_data
# [1] "foo"
```

A number of attributes are treated specially by R, see **?attributes** for full details. Some of the special attributes will now be explained.

B.2.4.1 Object class and `class` attribute

Every object in R "has a class", meaning that **class(obj)** returns a character vector with the class of **obj**. Some objects have an *implicit* class, such as basic vectors

```
class(1:3)
# [1] "integer"
class(c(TRUE, FALSE))
# [1] "logical"
class(c("TRUE", "FALSE"))
# [1] "character"
```

but we can also set the class explicitly, either by using **attr** or by using **class** in the left-hand side of an expression:

```
a <- 1:3
class(a) <- "foo"
a
# [1] 1 2 3
# attr(,"class")
# [1] "foo"
class(a)
# [1] "foo"
attributes(a)
```

```
# $class
# [1] "foo"
```

in which case the newly set class overrides the earlier implicit class. This way, we can add methods for class **foo** by appending the class name to the method name:

```
print.foo <- function(x, ...) {
    print(paste("an object of class foo with length", length(x)))
}
print(a)
# [1] "an object of class foo with length 3"
```

Providing such methods is generally intended to create more usable software, but at the same time they may make the objects more opaque. It is sometimes useful to see what an object "is made of" by printing it after the class attribute is removed, as in

```
unclass(a)
# [1] 1 2 3
```

As a more elaborate example, consider the case where a polygon is made using package **sf**:

```
library(sf) |> suppressPackageStartupMessages()
p <- st_polygon(list(rbind(c(0,0), c(1,0), c(1,1), c(0,0))))
p
# POLYGON ((0 0, 1 0, 1 1, 0 0))
```

which prints the well-known-text form; to understand what the data structure is like, we can use

```
unclass(p)
# [[1]]
#      [,1] [,2]
# [1,]   0    0
# [2,]   1    0
# [3,]   1    1
# [4,]   0    0
```

B.2.4.2   The dim attribute

The **dim** attribute sets the matrix or array dimensions:

```
a <- 1:8
class(a)
# [1] "integer"
```

```
attr(a, "dim") <- c(2,4) # or: dim(a) = c(2,4)
class(a)
# [1] "matrix" "array"
a
#         [,1] [,2] [,3] [,4]
# [1,]     1    3    5    7
# [2,]     2    4    6    8
attr(a, "dim") <- c(2,2,2) # or: dim(a) = c(2,2,2)
class(a)
# [1] "array"
a
# , , 1
#
#         [,1] [,2]
# [1,]     1    3
# [2,]     2    4
#
# , , 2
#
#         [,1] [,2]
# [1,]     5    7
# [2,]     6    8
```

B.2.5   The names attributes

Named vectors carry their names in a **names** attribute. We saw examples for lists
above, an example for a numeric vector is:

```
a <- c(first = 3, second = 4, last = 5)
a["second"]
# second
#      4
attributes(a)
# $names
# [1] "first"  "second" "last"
```

Other name attributes include **dimnames** for **matrix** or **array**, which not only names
dimensions but also the labels associated values of each of the dimensions:

```
a <- matrix(1:4, 2, 2)
dimnames(a) <- list(rows = c("row1", "row2"),
                    cols = c("col1", "col2"))
a
#        cols
# rows    col1 col2
#   row1    1    3
```

```
#   row2    2    4
attributes(a)
# $dim
# [1] 2 2
#
# $dimnames
# $dimnames$rows
# [1] "row1" "row2"
#
# $dimnames$cols
# [1] "col1" "col2"
```

Data.frame objects have rows and columns, and each has names:

```
df <- data.frame(a = 1:3, b = c(TRUE, FALSE, TRUE))
attributes(df)
# $names
# [1] "a" "b"
#
# $class
# [1] "data.frame"
#
# $row.names
# [1] 1 2 3
```

B.2.6   Using structure

When programming, the pattern of adding or modifying attributes before returning an object is extremely common, an example being:

```
f <- function(x) {
   a <- create_obj(x) # call some other function
   attributes(a) <- list(class = "foo", meta = 33)
   a
}
```

The last two statements can be contracted in

```
f <- function(x) {
   a <- create_obj(x) # call some other function
   structure(a, class = "foo", meta = 33)
}
```

where function **structure** adds, replaces, or (in case of value NULL) removes attributes from the object in its first argument.

B.3 Dissecting a `MULTIPOLYGON`

We can use the above examples to dissect an `sf` object with `MULTIPOLYGON`s into pieces. Suppose we use the `nc` dataset,

```
system.file("gpkg/nc.gpkg", package = "sf") %>%
    read_sf() -> nc
```

we can see from the attributes of `nc`,

```
attributes(nc)
# $names
#  [1] "AREA"      "PERIMETER" "CNTY_"     "CNTY_ID"
#  [5] "NAME"      "FIPS"      "FIPSNO"    "CRESS_ID"
#  [9] "BIR74"     "SID74"     "NWBIR74"   "BIR79"
# [13] "SID79"     "NWBIR79"   "geom"
#
# $row.names
#   [1]   1   2   3   4   5   6   7   8   9  10  11  12  13  14
#  [15]  15  16  17  18  19  20  21  22  23  24  25  26  27  28
#  [29]  29  30  31  32  33  34  35  36  37  38  39  40  41  42
#  [43]  43  44  45  46  47  48  49  50  51  52  53  54  55  56
#  [57]  57  58  59  60  61  62  63  64  65  66  67  68  69  70
#  [71]  71  72  73  74  75  76  77  78  79  80  81  82  83  84
#  [85]  85  86  87  88  89  90  91  92  93  94  95  96  97  98
#  [99]  99 100
#
# $class
# [1] "sf"        "tbl_df"    "tbl"       "data.frame"
#
# $sf_column
# [1] "geom"
#
# $agr
#      AREA PERIMETER      CNTY_   CNTY_ID       NAME      FIPS
#      <NA>      <NA>       <NA>      <NA>       <NA>      <NA>
#    FIPSNO  CRESS_ID      BIR74     SID74    NWBIR74     BIR79
#      <NA>      <NA>       <NA>      <NA>       <NA>      <NA>
#     SID79   NWBIR79
#      <NA>      <NA>
# Levels: constant aggregate identity
```

that the geometry column is named **geom**. When we take out this column,

```
nc$geom
# Geometry set for 100 features
# Geometry type: MULTIPOLYGON
# Dimension:     XY
# Bounding box:  xmin: -84.3 ymin: 33.9 xmax: -75.5 ymax: 36.6
# Geodetic CRS:  NAD27
# First 5 geometries:
# MULTIPOLYGON (((-81.5 36.2, -81.5 36.3, -81.6 3...
# MULTIPOLYGON (((-81.2 36.4, -81.2 36.4, -81.3 3...
# MULTIPOLYGON (((-80.5 36.2, -80.5 36.3, -80.5 3...
# MULTIPOLYGON (((-76 36.3, -76 36.3, -76 36.3, -...
# MULTIPOLYGON (((-77.2 36.2, -77.2 36.2, -77.3 3...
```

we see an object that has the following attributes

```
attributes(nc$geom)
# $n_empty
# [1] 0
#
# $crs
# Coordinate Reference System:
#   User input: NAD27
#   wkt:
# GEOGCRS["NAD27",
#     DATUM["North American Datum 1927",
#         ELLIPSOID["Clarke 1866",6378206.4,294.978698213898,
#             LENGTHUNIT["metre",1]]],
#     PRIMEM["Greenwich",0,
#         ANGLEUNIT["degree",0.0174532925199433]],
#     CS[ellipsoidal,2],
#         AXIS["geodetic latitude (Lat)",north,
#             ORDER[1],
#             ANGLEUNIT["degree",0.0174532925199433]],
#         AXIS["geodetic longitude (Lon)",east,
#             ORDER[2],
#             ANGLEUNIT["degree",0.0174532925199433]],
#     USAGE[
#         SCOPE["Geodesy."],
#         AREA["North and central America: Antigua and ..."],
#         BBOX[7.15,167.65,83.17,-47.74]],
#     ID["EPSG",4267]]
#
# $class
# [1] "sfc_MULTIPOLYGON" "sfc"
#
# $precision
# [1] 0
```

```
#
# $bbox
#   xmin  ymin  xmax  ymax
# -84.3  33.9 -75.5  36.6
```

When we take the *contents* of the fourth list element, we obtain

```
nc$geom[[4]]  |> format(width = 60, digits = 5)
# [1] "MULTIPOLYGON (((-76.009 36.32, -76.017 36.338, -76.033 36..."
```

which is a (classed) list,

```
typeof(nc$geom[[4]])
# [1] "list"
```

with attributes

```
attributes(nc$geom[[4]])
# $class
# [1] "XY"              "MULTIPOLYGON" "sfg"
```

and length

```
length(nc$geom[[4]])
# [1] 3
```

The length indicates the number of outer rings: a multi-polygon can consist of more than one polygon. We see that most counties only have a single polygon:

```
lengths(nc$geom)
#    [1] 1 1 1 3 1 1 1 1 1 1 1 1 1 1 1 1 1 1 1 1 1 1 1 1 1 1 1 1
#   [29] 1 1 1 1 1 1 1 1 1 1 1 1 1 1 1 1 1 1 1 1 1 1 1 1 1 1 1 3
#   [57] 2 1 1 1 1 1 1 1 1 1 1 1 1 1 1 1 1 1 1 1 1 1 1 1 1 1 1 1
#   [85] 1 1 2 1 1 1 2 1 1 1 2 1 1 1 1 1
```

A multi-polygon is a list with polygons,

```
typeof(nc$geom[[4]])
# [1] "list"
```

and the *first* polygon of the fourth multi-polygon is again a list, because polygons have an outer ring *possibly* followed by multiple inner rings (holes)

—

```
typeof(nc$geom[[4]][[1]])
# [1] "list"
```

we see that it contains only one ring, the exterior ring:

```
length(nc$geom[[4]][[1]])
# [1] 1
```

and we can print type, the dimension and the first set of coordinates by

```
typeof(nc$geom[[4]][[1]][[1]])
# [1] "double"
dim(nc$geom[[4]][[1]][[1]])
# [1] 26  2
head(nc$geom[[4]][[1]][[1]])
#        [,1] [,2]
# [1,] -76.0 36.3
# [2,] -76.0 36.3
# [3,] -76.0 36.3
# [4,] -76.0 36.4
# [5,] -76.1 36.3
# [6,] -76.2 36.4
```

and we can now for instance change the latitude of the third coordinate by

```
nc$geom[[4]][[1]][[1]][3,2] <- 36.5
```

References

Alam, Moudud, Lars Ronnegard, and Xia Shen. 2019. *Hglm: Hierarchical Generalized Linear Models*. https://CRAN.R-project.org/package=hglm.

Alam, Moudud, Lars Rönnegård, and Xia Shen. 2015. "Fitting Conditional and Simultaneous Autoregressive Spatial Models in Hglm." *The R Journal* 7 (2): 5–18. https://doi.org/10.32614/RJ-2015-017.

Anselin, Luc. 1988. *Spatial Econometrics: Methods and Models*. Kluwer Academic Publishers.

———. 1995. "Local indicators of spatial association - LISA." *Geographical Analysis* 27 (2): 93–115.

———. 1996. "The Moran Scatterplot as an ESDA Tool to Assess Local Instability in Spatial Association." In *Spatial Analytical Perspectives on GIS*, edited by M. M. Fischer, H. J. Scholten, and D. Unwin, 111–25. London: Taylor & Francis.

———. 2019. "A Local Indicator of Multivariate Spatial Association: Extending Geary's c." *Geographical Analysis* 51 (2): 133–50. https://doi.org/10.1111/gean.12164.

Anselin, Luc, Xun Li, and Julia Koschinsky. 2021. "GeoDa, from the Desktop to an Ecosystem for Exploring Spatial Data." *Geographical Analysis*. https://doi.org/10.1111/gean.12311.

Appel, Marius. 2023. *Gdalcubes: Earth Observation Data Cubes from Satellite Image Collections*. https://github.com/appelmar/gdalcubes_R.

Appel, Marius, and Edzer Pebesma. 2019. "On-Demand Processing of Data Cubes from Satellite Image Collections with the Gdalcubes Library." *Data* 4 (3): 92. https://www.mdpi.com/2306-5729/4/3/92.

Appel, Marius, Edzer Pebesma, and Matthias Mohr. 2021. *Cloud-Based Processing of Satellite Image Collections in R Using STAC, COGs, and on-Demand Data Cubes*. https://r-spatial.org/r/2021/04/23/cloud-based-cubes.html.

Appelhans, Tim, Florian Detsch, Christoph Reudenbach, and Stefan Woellauer. 2022. *Mapview: Interactive Viewing of Spatial Data in r*. https://github.com/r-spatial/mapview.

Assunção, R. M., and E. A. Reis. 1999. "A New Proposal to Adjust Moran's I for Population Density." *Statistics in Medicine* 18: 2147–62.

Avis, D., and J. Horton. 1985. "Remarks on the Sphere of Influence Graph." In *Discrete Geometry and Convexity*, edited by J. E. Goodman, 323–27. New York: New York Academy of Sciences, New York.

Aybar, Cesar. 2022. *Rgee: R Bindings for Calling the Earth Engine API*. https://CRAN.R-project.org/package=rgee.

Baddeley, Adrian, Ege Rubak, and Rolf Turner. 2015. *Spatial Point Patterns: Methodology and Applications with R*. Chapman & Hall/CRC.

Baddeley, Adrian, Rolf Turner, and Ege Rubak. 2022. *Spatstat: Spatial Point Pattern Analysis, Model-Fitting, Simulation, Tests*. http://spatstat.org/.

Bates, Douglas, Martin Maechler, Ben Bolker, and Steven Walker. 2022. *Lme4: Linear Mixed-Effects Models Using Eigen and S4.* https://github.com/lme4/lme4/.

Bates, Douglas, Martin Maechler, and Mikael Jagan. 2022. *Matrix: Sparse and Dense Matrix Classes and Methods.* https://CRAN.R-project.org/package=Matrix.

Baumann, Peter, Eric Hirschorn, and Joan Masó. 2017. "OGC Coverage Implementation Schema." *OGC Implementation Standard.* https://docs.opengeospatial.or g/is/09-146r6/09-146r6.html.

Bavaud, F. 1998. "Models for Spatial Weights: A Systematic Look." *Geographical Analysis* 30: 153–71. https://doi.org/10.1111/j.1538-4632.1998.tb00394.x.

Becker, Marc, and Patrick Schratz. 2022. *Mlr3spatial: Support for Spatial Objects Within the Mlr3 Ecosystem.* https://CRAN.R-project.org/package=mlr3spatial.

Benjamin, Daniel J., James O. Berger, Johannesson Magnus, Brian A. Nosek, Wagenmakers E-J, Richard Berk, Kenneth A. Bollen, et al. 2018. "Redefine Statistical Significance." *Nature Human Behaviour* 2 (1): 6–10.

Benjamini, Yoav, and Yosef Hochberg. 1995. "Controlling the False Discovery Rate: A Practical and Powerful Approach to Multiple Testing." *Journal of the Royal Statistical Society. Series B (Methodological)* 57 (1): 289–300. https://doi.org/10 .1111/j.2517-6161.1995.tb02031.x.

Benjamini, Yoav, and Daniel Yekutieli. 2001. "The control of the false discovery rate in multiple testing under dependency." *The Annals of Statistics* 29 (4): 1165–88. https://doi.org/10.1214/aos/1013699998.

Besag, Julian. 1974. "Spatial Interaction and the Statistical Analysis of Lattice Systems." *Journal of the Royal Statistical Society. Series B (Methodological)* 36: pp. 192–236.

BIPM, IEC, ILAC IFCC, IUPAP IUPAC, and OIML ISO. 2012. "The International Vocabulary of Metrology–Basic and General Concepts and Associated Terms (VIM), 3rd Edn. JCGM 200: 2012." *JCGM (Joint Committee for Guides in Metrology).* https://www.bipm.org/en/publications/guides/.

Bivand, Roger. 2002. "Spatial Econometrics Functions in R: Classes and Methods." *Journal of Geographical Systems* 4: 405–21.

———. 2012. "After 'Raising the Bar': Applied Maximum Likelihood Estimation of Families of Models in Spatial Econometrics." *Estadística Española* 54: 71–88.

———. 2017. "Revisiting the Boston Data Set — Changing the Units of Observation Affects Estimated Willingness to Pay for Clean Air." *REGION* 4 (1): 109–27. https://doi.org/10.18335/region.v4i1.107.

———. 2020. *Why Have CRS, Projections and Transformations Changed?* https://rgdal.r-forge.r-project.org/articles/CRS_projections_ transformations.html.

———. 2022a. *classInt: Choose Univariate Class Intervals.* https://CRAN.R-project. org/package=classInt.

———. 2022b. "R Packages for Analyzing Spatial Data: A Comparative Case Study with Areal Data." *Geographical Analysis* 54 (3): 488–518. https://doi.org/10.111 1/gean.12319.

———. 2022c. *Spdep: Spatial Dependence: Weighting Schemes, Statistics.*

Bivand, Roger, and Virgilio Gómez-Rubio. 2021. "Spatial Survival Modelling of Business Re-Opening After Katrina: Survival Modelling Compared to Spatial Probit Modelling of Re-Opening Within 3, 6 or 12 Months." *Statistical Modelling* 21 (1-2): 137–60. https://doi.org/10.1177/1471082X20967158.

Bivand, Roger, Virgilio Gómez-Rubio, and Håvard Rue. 2015. "Spatial Data Analysis with r-INLA with Some Extensions." *Journal of Statistical Software, Articles* 63 (20): 1–31. https://doi.org/10.18637/jss.v063.i20.

Bivand, Roger, Giovanni Millo, and Gianfranco Piras. 2021. "A Review of Software for Spatial Econometrics in R." *Mathematics* 9 (11). https://doi.org/10.3390/math9111276.

Bivand, Roger, W. Müller, and M. Reder. 2009. "Power Calculations for Global and Local Moran's *I*." *Computational Statistics and Data Analysis* 53: 2859–72.

Bivand, Roger, Jakub Nowosad, and Robin Lovelace. 2022. *spData: Datasets for Spatial Analysis*. https://jakubnowosad.com/spData/.

Bivand, Roger, Edzer Pebesma, and Virgilio Gómez-Rubio. 2013. *Applied Spatial Data Analysis with R, Second Edition*. Springer, NY. http://www.asdar-book.org/.

Bivand, Roger, and Gianfranco Piras. 2015. "Comparing Implementations of Estimation Methods for Spatial Econometrics." *Journal of Statistical Software* 63 (1): 1–36. https://doi.org/10.18637/jss.v063.i18.

———. 2022. *Spatialreg: Spatial Regression Analysis*. https://CRAN.R-project.org/package=spatialreg.

Bivand, Roger, and B. A. Portnov. 2004. "Exploring Spatial Data Analysis Techniques Using R: The Case of Observations with No Neighbours." In *Advances in Spatial Econometrics: Methodology, Tools, Applications*, edited by Luc Anselin, Raymond J. G. M. Florax, and S. J. Rey, 121–42. Berlin: Springer.

Bivand, Roger, Zhe Sha, Liv Osland, and Ingrid Sandvig Thorsen. 2017. "A Comparison of Estimation Methods for Multilevel Models of Spatially Structured Data." *Spatial Statistics*. https://doi.org/10.1016/j.spasta.2017.01.002.

Bivand, Roger, and David W. S. Wong. 2018. "Comparing Implementations of Global and Local Indicators of Spatial Association." *TEST* 27 (3): 716–48. https://doi.org/10.1007/s11749-018-0599-x.

Blangiardo, Marta, and Michela Cameletti. 2015. *Spatial and Spatio-Temporal Bayesian Models with r-INLA*. John Wiley & Sons.

Boots, B., and A. Okabe. 2007. "Local Statistical Spatial Analysis: Inventory and Prospect." *International Journal of Geographical Information Science* 21 (4): 355–75. https://doi.org/10.1080/13658810601034267.

Breidt, F Jay, Jean D Opsomer, et al. 2017. "Model-Assisted Survey Estimation with Modern Prediction Techniques." *Statistical Science* 32 (2): 190–205.

Brody, Howard, Michael Russell Rip, Peter Vinten-Johansen, Nigel Paneth, and Stephen Rachman. 2000. "Map-Making and Myth-Making in Broad Street: The London Cholera Epidemic, 1854." *The Lancet* 356 (9223): 64–68. https://doi.org/10.1016/S0140-6736(00)02442-9.

Brooks, Mollie E., Kasper Kristensen, Koen J. van Benthem, Arni Magnusson, Casper W. Berg, Anders Nielsen, Hans J. Skaug, Martin Maechler, and Benjamin M. Bolker. 2017. "glmmTMB Balances Speed and Flexibility Among Packages for Zero-Inflated Generalized Linear Mixed Modeling." *The R Journal* 9 (2): 378–400. https://journal.r-project.org/archive/2017/RJ-2017-066/index.html.

Brown, C. F., Steven P Brumby, Brookie Guzder-Williams, Tanya Birch, Samantha Brooks Hyde, Joseph Mazzariello, Wanda Czerwinski, et al. 2022. "Dynamic World, Near Real-Time Global 10 m Land Use Land Cover Mapping." *Scientific Data* 9 (1): 1–17. https://doi.org/10.1038/s41597-022-01307-4.

Brown, P. G. 2010. "Overview of SciDB: Large Scale Array Storage, Processing and Analysis." In *Proceedings of the 2010 ACM SIGMOD International Conference on Management of Data*, 963–68. ACM.

Brus, Dick J. 2021a. "Statistical Approaches for Spatial Sample Survey: Persistent Misconceptions and New Developments." *European Journal of Soil Science* 72 (2): 686–703. https://doi.org/10.1111/ejss.12988.

———. 2021b. "Statistical Approaches for Spatial Sample Survey: Persistent Misconceptions and New Developments." *European Journal of Soil Science* 72 (2): 686–703. https://doi.org/10.1111/ejss.12988.

Bureau International des Poids et Mesures. 2006. *The International System of Units (SI), 8th Edition*. Organisation Intergouvernementale de la Convention du Mètre. https://www.bipm.org/en/publications/si-brochure/download.html.

Butler, H., M. Daly, A. Doyl, S. Gillies, S. Hagen, and T. Schaub. 2016. "The GeoJSON Format." Vol. Request for Comments: 7946. Internet Engineering Task Force (IETF). https://tools.ietf.org/html/rfc7946.

Caldas de Castro, Marcia, and Burton H. Singer. 2006. "Controlling the False Discovery Rate: A New Application to Account for Multiple and Dependent Tests in Local Statistics of Spatial Association." *Geographical Analysis* 38 (2): 180–208. https://doi.org/10.1111/j.0016-7363.2006.00682.x.

Chambers, John. 2016. *Extending R*. CRC Press.

Chrisman, Nicholas. 2012. "A Deflationary Approach to Fundamental Principles in GIScience." In *Francis Harvey (Ed.) Are There Fundamental Principles in Geographic Information Science?*, 42–64. CreateSpace, United States.

Clementini, Eliseo, Paolino Di Felice, and Peter van Oosterom. 1993. "A Small Set of Formal Topological Relationships Suitable for End-User Interaction." In *Advances in Spatial Databases*, edited by David Abel and Beng Chin Ooi, 277–95. Berlin, Heidelberg: Springer Berlin Heidelberg.

Cliff, A. D., and J. K. Ord. 1972. "Testing for Spatial Autocorrelation Among Regression Residuals." *Geographical Analysis* 4: 267–84.

———. 1973. *Spatial Autocorrelation*. London: Pion.

———. 1981. *Spatial Processes*. London: Pion.

Cobb, George W., and David S. Moore. 1997. "Mathematics, Statistics and Teaching." *The American Mathematical Monthly* 104: 801–23. https://www.jstor.org/stable/2975286.

Collins, Sarah N., Robert S. James, Pallav Ray, Katherine Chen, Angie Lassman, and James Brownlee. 2013. "Grids in Numerical Weather and Climate Models." In *Climate Change and Regional/Local Responses*, edited by Yuanzhi Zhang and Pallav Ray. Rijeka: IntechOpen. https://doi.org/10.5772/55922.

Cressie, N. A. C. 1993. *Statistics for Spatial Data*. New York: Wiley.

Csardi, Gabor, and Tamas Nepusz. 2006. "The Igraph Software Package for Complex Network Research." *InterJournal* Complex Systems: 1695. https://igraph.org.

Davies, Tilman, and David Bryant. 2013. "On Circulant Embedding for Gaussian Random Fields in R." *Journal of Statistical Software, Articles* 55 (9): 1–21. https://doi.org/10.18637/jss.v055.i09.

De Gruijter, J. J., Dick J. Brus, Marc F. P. Bierkens, and Martin Knotters. 2006. *Sampling for Natural Resource Monitoring*. Springer Science & Business Media.

De Gruijter, J. J., and C. J. F. Ter Braak. 1990. "Model-Free Estimation from Spatial Samples: A Reappraisal of Classical Sampling Theory." *Mathematical Geology* 22 (4): 407–15.

Diggle, P. J., and P. J. Ribeiro Jr. 2007. *Model-Based Geostatistics*. New York: Springer.

Diggle, P. J., J. A. Tawn, and R. A. Moyeed. 1998. "Model-Based Geostatistics." *Applied Statistics*, 299–350.

Do, Van Huyen, Thibault Laurent, and Anne Vanhems. 2021. "Guidelines on Areal Interpolation Methods." In *Advances in Contemporary Statistics and Econometrics: Festschrift in Honor of Christine Thomas-Agnan*, edited by Abdelaati Daouia and Anne Ruiz-Gazen, 385–407. Cham: Springer International Publishing. https://doi.org/10.1007/978-3-030-73249-3_20.

Do, Van Huyen, Christine Thomas-Agnan, and Anne Vanhems. 2015a. "Accuracy of Areal Interpolation Methods for Count Data." *Spatial Statistics* 14: 412–38. https://doi.org/10.1016/j.spasta.2015.07.005.

———. 2015b. "Spatial Reallocation of Areal Data: A Review." *Rev. Econ. Rég. Urbaine* 1/2: 27–58. https://www.tse-fr.eu/sites/default/files/medias/doc/wp/mad/wp_tse_397_v2.pdf.

Duncan, O. D., R. P. Cuzzort, and B. Duncan. 1961. *Statistical Geography: Problems in Analyzing Areal Data*. Glencoe, IL: Free Press.

Dunnington, Dewey. 2022. *Ggspatial: Spatial Data Framework for Ggplot2*. https://CRAN.R-project.org/package=ggspatial.

Dunnington, Dewey, Edzer Pebesma, and Ege Rubak. 2023. *S2: Spherical Geometry Operators Using the S2 Geometry Library*. https://CRAN.R-project.org/package=s2.

Eaton, Brian, Jonathan Gregory, Bob Drach, Karl Taylor, Steve Hankin, Jon Blower, John Caron, et al. 2022. *NetCDF Climate and Forecast (CF) Metadata Conventions, Version 1.10*. https://cfconventions.org/.

Eddelbuettel, Dirk. 2013. *Seamless R and C++ Integration with Rcpp*. Springer.

Egenhofer, Max J., and Robert D. Franzosa. 1991. "Point-Set Topological Spatial Relations." *International Journal of Geographical Information Systems* 5 (2): 161–74. https://doi.org/10.1080/02693799108927841.

Elhorst, J. Paul. 2010. "Applied Spatial Econometrics: Raising the Bar." *Spatial Economic Analysis* 5: 9–28.

Evenden, Gerald I. 1990. *Cartographic Projection Procedures for the UNIX Environment — a User's Manual*. http://download.osgeo.org/proj/OF90-284.pdf.

Evers, Kristian, and Thomas Knudsen. 2017. *Transformation Pipelines for PROJ.4*. https://www.fig.net/resources/proceedings/fig_proceedings/fig2017/papers/iss6b/ISS6B_evers_knudsen_9156.pdf.

Fellows, Ian, and using the JMapViewer library by Jan Peter Stotz. 2019. *OpenStreetMap: Access to Open Street Map Raster Images*. https://CRAN.R-project.org/package=OpenStreetMap.

Fingleton, B. 1999. "Spurious spatial regression: Some Monte Carlo results with a spatial unit root and spatial cointegration." *Journal of Regional Science* 9: 1–19.

Florax, Raymond J. G. M., Hendrik Folmer, and Sergio J. Rey. 2006. "A Comment on Specification Searches in Spatial Econometrics: The Relevance of Hendry's Methodology: A Reply." *Regional Science and Urban Economics* 36 (2): 300–308. https://doi.org/10.1016/j.regsciurbeco.2005.10.002.

Florax, Raymond J. G. M, Hendrik Folmer, and Sergio J Rey. 2003. "Specification Searches in Spatial Econometrics: The Relevance of Hendry's Methodology." *Regional Science and Urban Economics* 33 (5): 557–79. https://doi.org/10.1016/S0166-0462(03)00002-4.

Freni-Sterrantino, Anna, Massimo Ventrucci, and Håvard Rue. 2018. "A Note on Intrinsic Conditional Autoregressive Models for Disconnected Graphs." *Spatial and Spatio-Temporal Epidemiology* 26: 25–34. https://doi.org/10.1016/j.sste.2018.04.002.

Gabriel, Edith, Peter J Diggle, Barry Rowlingson, and Francisco J Rodriguez-Cortes. 2022. *Stpp: Space-Time Point Pattern Simulation, Visualisation and Analysis.* https://CRAN.R-project.org/package=stpp.

Gabriel, Edith, Barry Rowlingson, and Peter Diggle. 2013. "Stpp: An R Package for Plotting, Simulating and Analyzing Spatio-Temporal Point Patterns." *Journal of Statistical Software, Articles* 53 (2): 1–29. https://doi.org/10.18637/jss.v053.i02.

Gaetan, Carlo, and Xavier Guyon. 2010. *Spatial Statistics and Modeling.* New York: Springer.

Galton, A. 2004. "Fields and Objects in Space, Time and Space-Time." *Spatial Cognition and Computation* 4.

Garnier, Simon. 2021. *Viridis: Colorblind-Friendly Color Maps for r.* https://CRAN.R-project.org/package=viridis.

Geary, R. C. 1954. "The Contiguity Ratio and Statistical Mapping." *The Incorporated Statistician* 5: 115–45.

Gerber, Florian, and Reinhard Furrer. 2015. "Pitfalls in the Implementation of Bayesian Hierarchical Modeling of Areal Count Data: An Illustration Using BYM and Leroux Models." *Journal of Statistical Software, Code Snippets* 63 (1): 1–32. https://doi.org/10.18637/jss.v063.c01.

Gerber, Florian, Rogier de Jong, Michael E Schaepman, Gabriela Schaepman-Strub, and Reinhard Furrer. 2018. "Predicting Missing Values in Spatio-Temporal Remote Sensing Data." *IEEE Transactions on Geoscience and Remote Sensing* 56 (5): 2841–53.

Getis, A., and J. K. Ord. 1992. "The Analysis of Spatial Association by the Use of Distance Statistics." *Geographical Analysis* 24 (2): 189–206.

———. 1996. "Local Spatial Statistics: An Overview." In *Spatial Analysis: Modelling in a GIS Environment,* edited by P. Longley and M Batty, 261–77. Cambridge: GeoInformation International.

Giraud, Timothée. 2022. *Mapsf: Thematic Cartography.* https://riatelab.github.io/mapsf/.

Gómez-Rubio, Virgilio. 2019. "Spatial Data Analysis with INLA. Coding Club UC3M Tutorial Series. Universidad Carlos III de Madrid." https://codingclubuc3m.rbind.io/talk/2019-11-05/.

———. 2020. *Bayesian Inference with INLA.* Boca Raton, FL: CRC Press.

Gómez-Rubio, Virgilio, Roger Bivand, and Håvard Rue. 2015. "A New Latent Class to Fit Spatial Econometrics Models with Integrated Nested Laplace Approximations." *Procedia Environmental Sciences* 27: 116–18. https://doi.org/10.1016/j.proenv.2015.07.119.

González, Álvaro. 2010. "Measurement of Areas on a Sphere Using Fibonacci and Latitude–Longitude Lattices." *Mathematical Geosciences* 42 (1): 49–64. https://arxiv.org/pdf/0912.4540.pdf.

Goodchild, Michael F, and Nina Siu Ngan Lam. 1980. *Areal Interpolation: A Variant of the Traditional Spatial Problem.* Department of Geography, University of Western Ontario London, ON, Canada.

Gorelick, Noel, Matt Hancher, Mike Dixon, Simon Ilyushchenko, David Thau, and Rebecca Moore. 2017. "Google Earth Engine: Planetary-Scale Geospatial Analysis for Everyone." *Remote Sensing of Environment* 202: 18–27. https://doi.org/10.1016/j.rse.2017.06.031.

Goulard, Michel, Thibault Laurent, and Christine Thomas-Agnan. 2017. "About Predictions in Spatial Autoregressive Models: Optimal and Almost Optimal Strategies." *Spatial Economic Analysis* 12 (2-3): 304–25. https://doi.org/10.1080/17421772.2017.1300679.

Gräler, Benedikt, Edzer Pebesma, and Gerard Heuvelink. 2016. "Spatio-Temporal Interpolation using gstat." *The R Journal* 8 (1): 204–18. https://doi.org/10.32614/RJ-2016-014.

Hahsler, Michael, and Matthew Piekenbrock. 2022. *Dbscan: Density-Based Spatial Clustering of Applications with Noise (DBSCAN) and Related Algorithms.* https://github.com/mhahsler/dbscan.

Halleck Vega, Solmaria, and J. Paul Elhorst. 2015. "The SLX Model." *Journal of Regional Science* 55 (3): 339–63. https://doi.org/10.1111/jors.12188.

Haltiner, G. J., and R. T. Williams. 1980. *Numerical Prediction and Dynamic Meteorology.* New York: John Wiley; Sons.

Hand, David J. 2004. *Measurement Theory and Practice.* Arnold, London.

Healy, Kieran. 2018. *Data Visualization, a Practical Introduction.* Princeton University Press. http://socviz.co/index.html.

Heaton, Matthew J., Abhirup Datta, Andrew O. Finley, Reinhard Furrer, Joseph Guinness, Rajarshi Guhaniyogi, Florian Gerber, et al. 2018. "A Case Study Competition Among Methods for Analyzing Large Spatial Data." *Journal of Agricultural, Biological and Environmental Statistics,* December. https://doi.org/10.1007/s13253-018-00348-w.

Hendry, David F. 2006. "A Comment on 'Specification Searches in Spatial Econometrics: The Relevance of Hendry's Methodology'." *Regional Science and Urban Economics* 36 (2): 309–12. https://doi.org/10.1016/j.regsciurbeco.2005.10.001.

Hepple, Leslie W. 1976. "A Maximum Likelihood Model for Econometric Estimation with Spatial Series." In *Theory and Practice in Regional Science,* edited by I. Masser, 90–104. London Papers in Regional Science. London: Pion.

Herring, J. R. 2010. "OpenGIS Implementation Standard for Geographic Information-Simple Feature Access-Part 2: SQL Option." *Open Geospatial Consortium Inc.* http://portal.opengeospatial.org/files/?artifact_id=25354.

———. 2011. "OpenGIS Implementation Standard for Geographic Information-Simple Feature Access-Part 1: Common Architecture." *Open Geospatial Consortium Inc,* 111. http://portal.opengeospatial.org/files/?artifact_id=25355.

Herring, J. R. et al. 2011. "Opengis® Implementation Standard for Geographic Information-Simple Feature Access-Part 1: Common Architecture [Corrigendum]."

Hersbach, Hans, Bill Bell, Paul Berrisford, Shoji Hirahara, András Horányi, Joaquín Muñoz-Sabater, Julien Nicolas, et al. 2020. "The ERA5 Global Reanalysis." *Quarterly Journal of the Royal Meteorological Society* 146 (730): 1999–2049. https://doi.org/10.1002/qj.3803.

Hijmans, Robert J. 2023a. *Raster: Geographic Data Analysis and Modeling.* https://rspatial.org/raster.

———. 2023b. *Terra: Spatial Data Analysis.* https://rspatial.org/terra/.

Hufkens, Koen. 2023. *Ecmwfr: Interface to ECMWF and CDS Data Web Services.* https://github.com/bluegreen-labs/ecmwfr.

Ihaka, Ross, Paul Murrell, Kurt Hornik, Jason C. Fisher, Reto Stauffer, Claus O. Wilke, Claire D. McWhite, and Achim Zeileis. 2023. *Colorspace: A Toolbox for*

Manipulating and Assessing Colors and Palettes. https://CRAN.R-project.org/package=colorspace.

Iliffe, Jonathan, and Roger Lott. 2008. *Datums and Map Projections for Remote Sensing, GIS, and Surveying. Whittles Pub.* CRC Press, Scotland, UK.

ISO. 2004. *Geographic Information – Simple Feature Access – Part 1: Common Architecture.*https://www.iso.org/standard/40114.html .

Jones, Philip W. 1999. "First- and Second-Order Conservative Remapping Schemes for Grids in Spherical Coordinates." *Mon. Wea. Rev.* 127: 2204–10. https://doi.org/10.1175/1520-0493(1999) .

Joo, Rocío, Matthew E. Boone, Thomas A. Clay, Samantha C. Patrick, Susana Clusella-Trullas, and Mathieu Basille. 2020. "Navigating Through the R Packages for Movement." *Journal of Animal Ecology* 89 (1): 248–67. https://doi.org/10.111 1/1365-2656.13116.

Journel, Andre G., and Charles J. Huijbregts. 1978. *Mining Geostatistics.* Academic Press, London.

Karney, Charles F. F. 2013. "Algorithms for Geodesics." *Journal of Geodesy* 87 (1): 43–55. https://link.springer.com/content/pdf/10.1007/s00190-012-0578-z.pdf.

Kelejian, Harry, and Gianfranco Piras. 2017. *Spatial Econometrics.* London: Academic Press.

Kitanidis, Peter K., and Robert W. Lane. 1985. "Maximum Likelihood Parameter Estimation of Hydrologic Spatial Processes by the Gauss-Newton Method." *Journal of Hydrology* 79 (1-2): 53–71. https://doi.org/10.1016/0022-1694(85)90181-7.

Knudsen, Thomas, and Kristian Evers. 2017. *Transformation Pipelines for PROJ.4.* https://meetingorganizer.copernicus.org/EGU2017/EGU2017-8050.pdf.

Kuhn, Max. 2022. *Caret: Classification and Regression Training.* https://github.com/topepo/caret/.

Kuhn, Max, and Hadley Wickham. 2022. *Tidymodels: Easily Install and Load the Tidymodels Packages.* https://CRAN.R-project.org/package=tidymodels.

Kyriakidis, P. C. 2004. "A Geostatistical Framework for Areal-to-Point Spatial Interpolation." *Geographical Analysis* 36: 259–89.

Laurent, Thibault, and Paula Margaretic. 2021. "Predictions in Spatial Econometric Models: Application to Unemployment Data." In *Advances in Contemporary Statistics and Econometrics: Festschrift in Honor of Christine Thomas-Agnan,* edited by Abdelaati Daouia and Anne Ruiz-Gazen, 409–26. Cham: Springer International Publishing. https://doi.org/10.1007/978-3-030-73249-3_21.

LeSage, James P. 2014. "What Regional Scientists Need to Know about Spatial Econometrics." *Review of Regional Studies* 44: 13–32. https://journal.srsa.org/ojs/index.php/RRS/article/view/44.1.2.

LeSage, James P., and Kelley R. Pace. 2009. *Introduction to Spatial Econometrics.* Boca Raton, FL: CRC Press.

Li, Xun, and Luc Anselin. 2021. *Rgeoda: R Library for Spatial Data Analysis.* https://CRAN.R-project.org/package=rgeoda.

———. 2022. *Rgeoda: R Library for Spatial Data Analysis.* https://CRAN.R-project.org/package=rgeoda.

Lott, Roger. 2015. "Geographic Information-Well-Known Text Representation of Coordinate Reference Systems." Open Geospatial Consortium.http://docs.opengeospatial.org/is/12-063r5/12-063r5.html .

Lovelace, Robin, Richard Ellison, and Malcolm Morgan. 2022. *Stplanr: Sustainable Transport Planning.* https://CRAN.R-project.org/package=stplanr.

Lovelace, Robin, Jakub Nowosad, and Jannes Muenchow. 2019. *Geocomputation with R*. Chapman & Hall/CRC.https://geocompr.robinlovelace.net/ .

Lu, Meng, Marius Appel, and Edzer Pebesma. 2018. "Multidimensional Arrays for Analysing Geoscientific Data." *ISPRS International Journal of Geo-Information* 7 (8): 313.

Lu, Meng, Edzer Pebesma, Alber Sanchez, and Jan Verbesselt. 2016. "Spatio-Temporal Change Detection from Multidimensional Arrays: Detecting Deforestation from MODIS Time Series." *ISPRS Journal of Photogrammetry and Remote Sensing* 117: 227–36. https://doi.org/10.1016/j.isprsjprs.2016.03.007.

Mark Padgham, Bob Rudis, Robin Lovelace, and Maëlle Salmon. 2017. *Osmdata. The Journal of Open Source Software*. Vol. 2. The Open Journal. https://doi.org/10.21105/joss.00305.

Martin, D. 1989. "Mapping Population Data from Zone Centroid Locations." *Transactions of the Institute of British Geographers, New Series* 14: 90–97.

Martinetti, Davide, and Ghislain Geniaux. 2017. "Approximate Likelihood Estimation of Spatial Probit Models." *Regional Science and Urban Economics* 64: 30–45. https://doi.org/10.1016/j.regsciurbeco.2017.02.002.

McCulloch, Charles E., and Shayle R. Searle. 2001. *Generalized, Linear, and Mixed Models*. New York: Wiley.

McMillen, Daniel P. 2003. "Spatial Autocorrelation or Model Misspecification?" *International Regional Science Review* 26: 208–17.

———. 2013. *Quantile Regression for Spatial Data*. Heidelberg: Springer-Verlag.

Mennis, Jeremy. 2003. "Generating Surface Models of Population Using Dasymetric Mapping." *The Professional Geographer* 55 (1): 31–42.

Meyer, Hanna, Carles Milà, and Marvin Ludwig. 2023. *CAST: Caret Applications for Spatial-Temporal Models*. https://CRAN.R-project.org/package=CAST.

Meyer, Hanna, and Edzer Pebesma. 2021. "Predicting into Unknown Space? Estimating the Area of Applicability of Spatial Prediction Models." *Methods in Ecology and Evolution* 12 (9): 1620–33. https://doi.org/10.1111/2041-210X.13650.

———. 2022. "Machine Learning-Based Global Maps of Ecological Variables and the Challenge of Assessing Them." *Nature Communincations* 13.https://doi.org/10.1038/s41467-022-29838-9 .

Mila, Carles, Jorge Mateu, Edzer Pebesma, and Hanna Meyer. 2022. "Nearest Neighbour Distance Matching Leave-One-Out Cross-Validation for Map Validation." *Methods in Ecology and Evolution* 13 (6): 1304–16. https://doi.org/10.1111/2041-210X.13851.

Militino, A. F., M. D. Ugarte, U. Pérez-Goya, and M. G. Genton. 2019. "Interpolation of the Mean Anomalies for Cloud-Filling in Land Surface Temperature and Normalized Difference Vegetation Index." *IEEE Transactions on Geoscience and Remote Sensing* 57 (8): 6068–78. https://doi.org/10.1109/TGRS.2019.2904193.

Millo, Giovanni, and Gianfranco Piras. 2012. "splm: Spatial Panel Data Models in R." *Journal of Statistical Software* 47 (1): 1–38.

Moran, P. A. P. 1948. "The Interpretation of Statistical Maps." *Journal of the Royal Statistical Society, Series B (Methodological)* 10 (2): 243–51.

Moreno, Mel, and Mathieu Basille. 2018. *Drawing Beautiful Maps Programmatically with r, Sf and Ggplot2 — Part 1: Basics*. https://www.r-spatial.org/r/2018/10/25/ggplot2-sf.html.

Mur, Jesús, and Ana Angulo. 2006. "The Spatial Durbin Model and the Common Factor Tests." *Spatial Economic Analysis* 1 (2): 207–26. https://doi.org/10.1080/17421770601009841.

Nepusz, Tamás. 2022. *Igraph: Network Analysis and Visualization.* https://CRAN.R-project.org/package=igraph.

Neuwirth, Erich. 2022. *RColorBrewer: ColorBrewer Palettes.* https://CRAN.R-project.org/package=RColorBrewer.

O'Brien, Joshua. 2022. *gdalUtilities: Wrappers for GDAL Utilities Executables.* https://github.com/JoshOBrien/gdalUtilities/.

Obe, Regina O., and Leo S. Hsu. 2015. *PostGIS in Action.* Manning Publications Co.

Okabe, A., T. Satoh, T. Furuta, A. Suzuki, and K. Okano. 2008. "Generalized Network Voronoi Diagrams: Concepts, Computational Methods, and Applications." *International Journal of Geographical Information Science* 22 (9): 965–94. https://doi.org/10.1080/13658810701587891.

Olsson, Gunnar. 1970. "Explanation, Prediction, and Meaning Variance: An Assessment of Distance Interaction Models." *Economic Geography* 46: 223–33. https://doi.org/10.2307/143140.

Ord, J. K. 1975. "Estimation Methods for Models of Spatial Interaction." *Journal of the American Statistical Association* 70 (349): 120–26.

Ord, J. K., and A. Getis. 2001. "Testing for Local Spatial Autocorrelation in the Presence of Global Autocorrelation." *Journal of Regional Science* 41 (3): 411–32.

Pace, R. K., and James P. LeSage. 2008. "A Spatial Hausman Test." *Economics Letters* 101: 282–84.

Papadopoulos, Stavros, Kushal Datta, Samuel Madden, and Timothy Mattson. 2016. "The Tiledb Array Data Storage Manager." *Proceedings of the VLDB Endowment* 10 (4): 349–60.

Pebesma, Edzer. 2004. "Multivariable Geostatistics in S: The Gstat Package." *Computers & Geosciences* 30: 683–91.

———. 2012. "spacetime: Spatio-Temporal Data in R." *Journal of Statistical Software* 51 (7): 1–30. https://www.jstatsoft.org/v51/i07/.

———. 2018. "Simple Features for R: Standardized Support for Spatial Vector Data." *The R Journal* 10 (1): 439–46. https://doi.org/10.32614/RJ-2018-009.

———. 2022a. *Reading Zarr Files with r Package Stars.* https://r-spatial.org/r/2022/09/13/zarr.html.

———. 2022b. *Sf: Simple Features for r.* https://CRAN.R-project.org/package=sf.

———. 2022c. *Spacetime: Classes and Methods for Spatio-Temporal Data.* https://github.com/edzer/spacetime.

———. 2022d. *Stars: Spatiotemporal Arrays, Raster and Vector Data Cubes.* https://CRAN.R-project.org/package=stars.

———. 2023. *Lwgeom: Bindings to Selected Liblwgeom Functions for Simple Features.* https://github.com/r-spatial/lwgeom/.

Pebesma, Edzer, and Benedikt Graeler. 2022. *Gstat: Spatial and Spatio-Temporal Geostatistical Modelling, Prediction and Simulation.* https://github.com/r-spatial/gstat/.

Pebesma, Edzer, Thomas Mailund, and James Hiebert. 2016. "Measurement Units in R." *The R Journal* 8 (2): 486–94. https://doi.org/10.32614/RJ-2016-061.

Pebesma, Edzer, Thomas Mailund, Tomasz Kalinowski, and Iñaki Ucar. 2022. *Units: Measurement Units for R Vectors.* https://github.com/r-quantities/units.

Pinheiro, Jose C., and Douglas M. Bates. 2000. *Mixed-Effects Models in S and S-Plus*. New York: Springer.

Piras, Gianfranco, and Ingmar R. Prucha. 2014. "On the Finite Sample Properties of Pre-Test Estimators of Spatial Models." *Regional Science and Urban Economics* 46: 103–15. https://doi.org/10.1016/j.regsciurbeco.2014.03.002.

Plate, Tony, and Richard Heiberger. 2016. *Abind: Combine Multidimensional Arrays*. https://CRAN.R-project.org/package=abind.

Ploton, Pierre, Frédéric Mortier, Maxime Réjou-Méchain, Nicolas Barbier, Nicolas Picard, Vivien Rossi, Carsten Dormann, et al. 2020. "Spatial Validation Reveals Poor Predictive Performance of Large-Scale Ecological Mapping Models." *Nature Communications* 11 (1): 4540. https://www.nature.com/articles/s41467-020-18321-y.

Raim, A. M., S. H. Holan, J. R. Bradley, and C. K. Wikle. 2021. "Spatio-Temporal Change of Support Modeling with r." *Computational Statistics* 36: 749–80. https://doi.org/ 10.1007/s00180-020-01029-4 .

Raoult, Baudouin, Cedric Bergeron, Angel López Alós, Jean-Noël Thépaut, and Dick Dee. 2017. "Climate Service Develops User-Friendly Data Store." *ECMWF Newsletter* 151: 22–27.

Ripley, B. D. 1981. *Spatial Statistics*. New York: Wiley.

———. 1988. *Statistical Inference for Spatial Processes*. Cambridge: Cambridge University Press.

Rue, Havard, Finn Lindgren, and Elias Teixeira Krainski. 2022. *INLA: Full Bayesian Analysis of Latent Gaussian Models Using Integrated Nested Laplace Approximations*.

Sarrias, Mauricio, and Gianfranco Piras. 2022. *Spldv: Spatial Models for Limited Dependent Variables*. https://CRAN.R-project.org/package=spldv.

Sauer, Jeffery, Taylor Oshan, Sergio Rey, and Levi John Wolf. 2021. "The Importance of Null Hypotheses: Understanding Differences in Local Moran's I_i Under Heteroskedasticity." *Geographical Analysis*. https://doi.org/10.1111/gean.12304.

Schabenberger, O., and C. A. Gotway. 2005. *Statistical Methods for Spatial Data Analysis*. Boca Raton/London: Chapman & Hall/CRC.

Scheider, Simon, Benedikt Gräler, Edzer Pebesma, and Christoph Stasch. 2016. "Modeling Spatiotemporal Information Generation." *International Journal of Geographical Information Science* 30 (10): 1980–2008. https://doi.org/10.1080/13 658816.2016.1151520.

Schlather, Martin. 2011. "Construction of Covariance Functions and Unconditional Simulation of Random Fields." In *Space-Time Processes and Challenges Related to Environmental Problems*, edited by E. Porcu, Montero J. M., and M. Schlather. New York: Springer.

Schlesinger, Thomas, and Manuel J. A. Eugster. 2013. *Osmar: OpenStreetMap and r*. http://osmar.r-forge.r-project.org/.

Schramm, Matthias, Edzer Pebesma, Milutin Milenković, Luca Foresta, Jeroen Dries, Alexander Jacob, Wolfgang Wagner, et al. 2021. "The openEO API– Harmonising the Use of Earth Observation Cloud Services Using Virtual Data Cube Functionalities." *Remote Sensing* 13 (6). https://doi.org/10.3390/rs130611 25.

Schratz, Patrick, and Marc Becker. 2022. *Mlr3spatiotempcv: Spatiotemporal Resampling Methods for Mlr3*. https://CRAN.R-project.org/package=mlr3spatiotemp cv.

She, Bing, Xinyan Zhu, Xinyue Ye, Wei Guo, Kehua Su, and Jay Lee. 2015. "Weighted Network Voronoi Diagrams for Local Spatial Analysis." *Computers, Environment and Urban Systems* 52: 70–80. https://doi.org/10.1016/j.compenvurbsys.2015.03.005.

Silge, Julia, and Michael Mahoney. 2023. *Spatialsample: Spatial Resampling Infrastructure.* https://CRAN.R-project.org/package=spatialsample.

Simoes, Rolf, Gilberto Camara, Gilberto Queiroz, Felipe Souza, Pedro R. Andrade, Lorena Santos, Alexandre Carvalho, and Karine Ferreira. 2021. "Satellite Image Time Series Analysis for Big Earth Observation Data." *Remote Sensing* 13 (13). https://doi.org/10.3390/rs13132428.

Simoes, Rolf, Felipe Carvalho, and Brazil Data Cube Team. 2023. *Rstac: Client Library for SpatioTemporal Asset Catalog.* https://brazil-data-cube.github.io/rstac/.

Skøien, Jon O, Günter Blöschl, Gregor Laaha, E Pebesma, Juraj Parajka, and Alberto Viglione. 2014. "Rtop: An R Package for Interpolation of Data with a Variable Spatial Support, with an Example from River Networks." *Computers & Geosciences* 67: 180–90.

Smith, Tony E. 2009. "Estimation Bias in Spatial Models with Strongly Connected Weight Matrices." *Geographical Analysis* 41 (3): 307–32. https://doi.org/10.1111/j.1538-4632.2009.00758.x.

Smith, Tony E., and K. L. Lee. 2012. "The effects of spatial autoregressive dependencies on inference in ordinary least squares: a geometric approach." *Journal of Geographical Systems* 14 (January): 91–124. https://doi.org/10.1007/s10109-011-0152-x.

Sokal, R. R, N. L. Oden, and B. A. Thomson. 1998. "Local Spatial Autocorrelation in a Biological Model." *Geographical Analysis* 30: 331–54.

Stasch, Christoph, Simon Scheider, Edzer Pebesma, and Werner Kuhn. 2014. "Meaningful Spatial Prediction and Aggregation." *Environmental Modelling & Software* 51: 149–65. https://doi.org/10.1016/j.envsoft.2013.09.006.

Stoyan, Dietrich, Francisco J. Rodríguez-Cortés, Jorge Mateu, and Wilfried Gille. 2017. "Mark Variograms for Spatio-Temporal Point Processes." *Spatial Statistics* 20: 125–47. https://doi.org/10.1016/j.spasta.2017.02.006.

Suesse, Thomas. 2018. "Marginal Maximum Likelihood Estimation of SAR Models with Missing Data." *Computational Statistics & Data Analysis* 120: 98–110. https://doi.org/10.1016/j.csda.2017.11.004.

Teickner, Henning, Edzer Pebesma, and Benedikt Graeler. 2022. *Sftime: Classes and Methods for Simple Feature Objects That Have a Time Column.* https://CRAN.R-project.org/package=sftime.

Tennekes, Martijn. 2018. "tmap: Thematic Maps in R." *Journal of Statistical Software* 84 (6): 1–39. https://doi.org/10.18637/jss.v084.i06.

———. 2022. *Tmap: Thematic Maps.* https://github.com/r-tmap/tmap.

Tiefelsdorf, M. 2002. "The Saddlepoint Approximation of Moran's I and Local Moran's I_i Reference Distributions and Their Numerical Evaluation." *Geographical Analysis* 34: 187–206.

Tobler, W. R. 1970. "A Computer Movie Simulating Urban Growth in the Detroit Region." *Economic Geography* 46: 234–40. https://doi.org/10.2307/143141.

———. 1979. "Smooth Pycnophylactic Interpolation for Geographical Regions." *Journal of the American Statistical Association* 74: 519–30.

UCAR. 2014. *UDUNITS 2.2.26 Manual*. https://www.unidata.ucar.edu/software/udunits/udunits-current/doc/udunits/udunits2.html.

———. 2020. *The NetCDF User's Guide*. https://www.unidata.ucar.edu/software/netcdf/docs/user_guide.html.

Umlauf, Nikolaus, Daniel Adler, Thomas Kneib, Stefan Lang, and Achim Zeileis. 2015. "Structured Additive Regression Models: An R Interface to BayesX." *Journal of Statistical Software* 63 (21): 1–46. http://www.jstatsoft.org/v63/i21/.

Umlauf, Nikolaus, Thomas Kneib, Stefan Lang, and Achim Zeileis. 2022. *R2BayesX: Estimate Structured Additive Regression Models with BayesX*. https://CRAN.R-project.org/package=R2BayesX.

Upton, G., and B. Fingleton. 1985. *Spatial Data Analysis by Example: Point Pattern and Qualitative Data*. New York: Wiley.

van der Meer, Lucas, Lorena Abad, Andrea Gilardi, and Robin Lovelace. 2022. *Sfnetworks: Tidy Geospatial Networks*. https://CRAN.R-project.org/package=sfnetworks.

Van Lieshout, M. N. M. 2019. *Theory of Spatial Statistics*. Boca Raton, FL: Chapman & Hall/CRC.

Veach, Eric, Jesse Rosenstock, Eric Engle, Robert Snedegar, Julien Basch, and Tom Manshreck. 2020. "S2 Geometry." *Website*. https://s2geometry.io/.

Ver Hoef, Jay M, and Noel Cressie. 1993. "Multivariable Spatial Prediction." *Mathematical Geology* 25 (2): 219–40.

Verbesselt, Jan, Rob Hyndman, Glenn Newnham, and Darius Culvenor. 2010. "Detecting Trend and Seasonal Changes in Satellite Image Time Series." *Remote Sensing of Environment* 114 (1): 106–15. https://doi.org/10.1016/j.rse.2009.08.014.

Vranckx, M., T. Neyens, and C. Faes. 2019. "Comparison of Different Software Implementations for Spatial Disease Mapping." *Spatial and Spatio-Temporal Epidemiology* 31: 100302. https://doi.org/10.1016/j.sste.2019.100302.

Wagner, Martin, and Achim Zeileis. 2019. "Heterogeneity and Spatial Dependence of Regional Growth in the EU: A Recursive Partitioning Approach." *German Economic Review* 20 (1): 67–82. https://doi.org/10.1111/geer.12146.

Wall, M. M. 2004. "A Close Look at the Spatial Structure Implied by the CAR and SAR Models." *Journal of Statistical Planning and Inference* 121: 311–24.

Waller, Lance A., and Carol A. Gotway. 2004. *Applied Spatial Statistics for Public Health Data*. Hoboken, NJ: John Wiley & Sons.

Wang, Earo, Di Cook, Rob Hyndman, and Mitchell O'Hara-Wild. 2022. *Tsibble: Tidy Temporal Data Frames and Tools*. https://tsibble.tidyverts.org.

Whittle, P. 1954. "On Stationary Processes in the Plane." *Biometrika* 41 (3-4): 434–49. https://doi.org/10.1093/biomet/41.3-4.434.

Wickham, Hadley. 2014a. *Advanced R, Second Edition*. CRC Press.https://adv-r.hadley.nz/ .

———. 2014b. "Tidy Data." *Journal of Statistical Software* 59 (1).https://www.jstatsoft.org/article/view/v059i10 .

———. 2016. *Ggplot2: Elegant Graphics for Data Analysis*. Springer.

———. 2022. *Tidyverse: Easily Install and Load the Tidyverse*. https://CRAN.R-project.org/package=tidyverse.

Wickham, Hadley, Mara Averick, Jennifer Bryan, Winston Chang, Lucy D'Agostino McGowan, Romain François, Garrett Grolemund, et al. 2019. "Welcome to the Tidyverse." *Journal of Open Source Software* 4 (43): 1686. https://joss.theoj.org/papers/10.21105/joss.01686.

Wickham, Hadley, Winston Chang, Lionel Henry, Thomas Lin Pedersen, Kohske Takahashi, Claus Wilke, Kara Woo, Hiroaki Yutani, and Dewey Dunnington. 2022. *Ggplot2: Create Elegant Data Visualisations Using the Grammar of Graphics.* https://CRAN.R-project.org/package=ggplot2.

Wickham, Hadley, and Garret Grolemund. 2017. *R for Data Science.* O'Reilly. http://r4ds.had.co.nz/.

Wikle, Christopher K, Andrew Zammit-Mangion, and Noel Cressie. 2019. *Spatio-Temporal Statistics with R.* CRC Press.

Wilhelm, Stefan, and Miguel Godinho de Matos. 2013. "Estimating Spatial Probit Models in R." *The R Journal* 5 (1): 130–43. https://doi.org/10.32614/RJ-2013-013.

Wilke, Claus O. 2019. *Fundamentals of Data Visualization.* O'Reilly Media, Inc. https://serialmentor.com/dataviz/.

Wood, S. N. 2017. *Generalized Additive Models: An Introduction with R.* 2nd ed. Chapman & Hall/CRC.

———. 2022. *Mgcv: Mixed GAM Computation Vehicle with Automatic Smoothness Estimation.* https://CRAN.R-project.org/package=mgcv.

Zeileis, Achim, Jason C. Fisher, Kurt Hornik, Ross Ihaka, Claire D. McWhite, Paul Murrell, Reto Stauffer, and Claus O. Wilke. 2020. "colorspace: A Toolbox for Manipulating and Assessing Colors and Palettes." *Journal of Statistical Software* 96 (1): 1–49. https://doi.org/10.18637/jss.v096.i01.

Index

aggregate
  data cube, 65
  sf, 81
aggregates, 51
aggregation
  spatial, 9
agr, attribute-geometry relationship,
  76
air quality aggregation, 99
altitude, 20
  direction, 20
antimeridian, 46
area-weighted interpolation, 10, 54,
  82
areal data, 191
array data, 59
arrow, 135
attribute-geometry relationship, 50
  aggregate, 50
  constant, 50
  identity, 51
attributes
  of R objects, 267
axis order, 25, 115
  disambiguate, 27

bbox
  attribute, 76
bounding box, 46
bounding cap, 46
bounding rectangle, 46

CAR models, 234
cdn.proj.org, 112
CIRCULARSTRING, 33
class, 268
cloud native, 131, 134
cloud optimised, 134
cloud-optimised GeoTIFF, 139

cokriging, 183
colour breaks, 122
COMPOUNDCURVE, 33
conditional autoregressive model, 233
conditional simulation, 173
conservative region aggregation, 54
conservative region regridding, 54
contiguous neighbours, 194
coordinate
  conversion, 23
  transformation, 23
coordinate reference systems, 7, 22
coordinate system, 23
coordinates
  altitude, 20
  as predictors, 152
  axis order, 25, 115
  bounding box, 46
  Cartesian, 19
  ellipsoidal, 8, 19
  geocentric, 19
  great circle, 45
  precisions, 39
  projected, 8, 21
  spherical or planar, 45
  straight line, 45
  units, 20
  WKT-2, 26
  z and m, 31
Copernicus, 140
coverage, 40
  tesselation, 41
crop raster dimensions, 88
cross-validation, 153
cross-variogram, 184
crs
  attribute, 76
CURVE, 33
CURVEPOLYGON, 33

curvilinear raster, 10, 98

dasymetric mapping, 55
data cube
    aggregate, 91
    aggregate to vector cube, 65
    apply function to dimension, 64
    attributes, 60
    dimension
        support, 62
    dimensions, 60
    extract, 91
    filter, 63
    machine learning with, 92
    origin-destination matrix, 102
    predictive models, 92
    raster dimensions, 62
    read with stars, 85
    reduce dimension, 64
    regular dimensions, 62
    support, 60
    switch dimension attribute, 67
    tidy, 108
    vector, 99
    vector geometry dimension, 62
data cubes, 12, 59
data science, xi, 5
    software, 13
data structures, 264
datum, 8, 18
    dynamic, 23
    grids, 25, 112
    transformation, 8
DE-9IM, 34
design-based inference, 151
dim, 269
dimension
    apply function to dimension, 64
    change order, 89
    permutation, 89
    reduce, 64
    regular, 62
    switch with attribute, 67
distance
    ellipsoidal, 22
    great circle, 21
    straight line, 21
distance-based neighbours, 199
downscaling, 56

dplyr, 75
dynamic spatial data, 67

ellipsoid of revolution, 20
ellipsoidal coordinates, 19
empty geometry, 32
encoding
    well-known binary, 33
    well-known text, 30
    WKB, 33
    WKT, 30
ERA5, 140, 142
extensive properties, 54
extract, 91

feature
    attributes, 6
fields, 51
footprint, spatial, 191
full polygon, 46

GDAL, 14
    barn raising, 24
    datum transformation, 24
    geolocation arrays, 62
    geotransform, 62
    multi-dimensional arrays, 140
    utils C API, 98
Geary's C, 215
    local, 226
general nested model, 246
geoarrow, 135
geodetic datum, 23
GeographicLib, 14
geoid, 23
geom_sf, 125
geom_stars, 126
geometry
    DE-9IM, 34
    empty, 32, 34
    measures
        binary, 37
        unary, 36
    operations, 34
    predicates, 34
        binary, 34
        unary, 34
    projected, 34
    relationship to attribute, 50

simple, 31, 34
simple feature, 29
support, 49
transformers
  binary, 38
  n-ary, 38
  unary, 37
valid, 31, 34
valid on the sphere, 47
well-known text, 30
WKT, 30
GEOMETRYCOLLECTION, 29
geopackage, 77
geoparquet, 135
GEOS, 14
GeoTIFF
cloud-optimised, 139
Getis-Ord G
local, 225
Getis-Org G, 215
ggplot2, 125
GNM, 246
Google Earth Engine, 136, 141
graticule, 123
great circle segment, 45

hglm, 238
hotspots, 220
hypercube, 59

impacts, 252
inference
design-based, 151
model-based, 151
INLA, 240, 255
intensive properties, 54
interpolation, 10, 165
area-to-area, 57
area-weighted, 10, 54, 82
inverse distance weighted, 167
kriging, 171
target grid, 166
intrinsic CAR model, 233
inverse distance weighted
interpolation, 167

joint-count test, 212

kernel density, 158
krige, 171

kriging
areal means, 171
block, 57, 171
cokriging, 183
conditional simulation, 173
multivariable, 183
ordinary, 171
spatiotemporal, 184
standard errors, 172
trend model, 174

Landsat, 140
large datasets, 131
latitude
geocentric, 20
geodetic, 20
lazy evaluation of computations, 137
leaflet, 128
mapview, 128
tmap, 128
liblwgeom, 14
linear model
conditional autoregressive, 233
intrinsic CAR, 233
mixed effects, 233
multilevel, 233
SAR, CAR, 234
LINESTRING, 29
list, 265
listw objects, 192
lme4, 238
longitude, 20

map, 6
map algebra, 93
maps
colour breaks, 122
colours, 121
graticule, 123
plotting, 119
plotting detail, 121
projections, 119
mapview, 128
Markov random field, 233
measurement units, 17
mgcv gam
ICAR, 241
model-based inference, 151
Moran's I, 210

approximations, 211
   empirical Bayes, 215
   global or local, 211
   hotspots, 220
   local, 218
   saddlepoint approximation, 222
MULTICURVE, 33
multilevel model, 233
multilevel models, 237
MULTILINESTRING, 29
MULTIPOINT, 29
MULTIPOLYGON, 29
MULTISURFACE, 33

nb objects, 192
nearest neighbours, 200
neighbourhood graph, 192
neighbours
   contiguous, 194
   distance-based, 199
   edge lengths, 197
   higher order, 206
   k nearest, 200
   queen, 195
   sphere of influence, 198
NetCDF, 14, 141
networks, 42
NULL, 267

objects, 51
observation window, 156
OGC:CRS84, 27
openEO, 141
openEO cloud, 136
OpenStreetMap, reading, 134
ordinary kriging, 171
origin-destination matrix, 102
out-of-core, 131

parquet, 135
pipes, 263
plot
   star, 93
POINT, 29
point pattern
   density model, 159
   homogeneity, 157
   line segments, 162
   marked, 161

   simulation, 163
   simulation on the sphere, 164
point pattern analysis, 155
point patterns, 10
Polish Presidential election data, 193
POLYGON, 29
polygon
   full vs. empty, 46
   inside or outside on sphere, 46
   inside, outside, 22
   on the sphere, 22
   ring direction, 46
   tesselation, 41
   valid on the sphere, 47
polygons, 6
POLYHEDRALSURFACE, 33
population density grid, 175
PostGIS, 14
ppm, 159
ppp, 156
precision
   attribute, 76
precisions, 39
predictive models, 152
PROJ, 14, 23
   datum transformation, 24
   pipelines, 113
   PROJ.4, 23
   proj4string, 23
   WKT-2, 25
proj.db, 112
projection, 21
   accuracy, 25
   properties, 120
   rasters, 116
projections, 7
proximity, areal data, 191

quadrat.test, 157
quantities, 17
queen neighbours, 195

R2BayesX, 239
random sampling, 151
raster, 84, 260
   curvilinear, 10, 62
   rectilinear, 10, 62
   regular, 10
   resolution, 136

rotated, 10
sheared, 10
tesselation, 41
raster data, 9
raster data cube, 62
raster-to-vector, 10
rasters
    warp, 116
Rcpp, 75
read_mdim, 141
read_stars
    proxy objects, 137
rectilinear raster, 10
reduce dimension, 64
regression
    generalised least squares, 233
    spatial, 233
resolution
    low-resolution computation, 136
rgdal, 75
    retirement, 259
rgeoda, 209, 229
rgeos, 75
    retirement, 259

s2geometry, 14
SAC, 246
sample variogram, 167
SAR models, 234
SARAR, 246
SDEM, 246
SDM, 246
SEM, 245
Sentinel Hub, 136
sf, 75
    aggregate, 81
    as.ppp, 156
    components figure, 76
    coordinate reference system, 78
    create from scratch, 76
    dbplyr proxy, 133
    delete a layer, 78
    filter, 78
    introduction, 75
    large datasets, 132
    migration from sp, 260
    plot
        legend, 123
    plot method, 123

projection, 110
read from file, 77
select using predicates, 78
spatial join, 81
subsetting, 78
update, append to a layer, 78
write to file, 78
sf_read
    from database connection, 133
sfc, class, 75
sfg, class, 75
sheared raster, 10
simple feature
    definition, 29
    geometry, 29
simple feature access, 29
simple feature geometry, 29
    types, 29
    z and m, 31
simple geometry, 31
SLX, 246
sp, 75
    differences from sf, 259
space
    bounded, 22
    unbounded, 22
spacetime, 84
spatial autocorrelation
    local measures of, 216
    measures of, 209
    of residuals, 211
spatial autocorrelation in fields, 165
spatial autocorrelation on graphs, 192
spatial autoregressive-autoregressive
        model, 246
spatial cross-validation, 153
spatial data science, xi
    software, 13
spatial Durbin error model, 246
spatial Durbin model, 246
spatial econometrics models, 245
spatial error model, SEM, 245
spatial graphs, disconnected, 192
spatial interpolation, 165
spatial random sampling, 151
spatial regression, 233
spatial weights matrix, 209
spatially lagged covariates, SAC, 246
spatially lagged X model, SLX, 246

spatialreg, 194
spatstat, 155
spDataLarge, 193
spdep, 194
sphere, 22
sphere of influence, 198
SQL-MM part 3, 33
st_as_stars, 139
    for generating a grid, 166
st_crop
    stars_proxy, 139
st_extract, 176
st_read
    geoparquet, 135
    inside bounding box, 132
    online resources, 134
    OpenStreetMap, 134
    query argument, 132
    vsizip, 132
st_read geoarrow, 135
stars, 84
    apply, 96
    arithmetic ops, 93
    crop, 88
    data cube, 84
    filter, 86
    lazy evaluation, 137
    map algebra, 93
    plot, 93
    projection, 110
    proxy objects, 137
    reading data cube, 85
    redimension, 89
    subset, 86
    warp, 116
stars_proxy
    lazy operations, 139
    st_as_stars, 139
statistical models, 147
superpopulation model, 151
support, 12, 49
    area, 12
    block, 12, 50
    data cube, 62
    dimension, 62
    downscaling, 50
    in file formats, 55
    point, 12, 50
    raster cells, 12

    temporal, 62
    upscaling, 50
SURFACE, 33

terra, 84, 260
tesselation, 41
    time, 41
tidyr, 75
tidyverse, 75
time series, 12
TIN, 33
tmap, 127, 193
    interactive views, 128
topology, 40
trajectories, 67
TRIANGLE, 33

udunits2, 14
units, 17
upscaling, 56

valid geometry, 31
variogram
    cross, 184
    model, 169
        maximum likelihood
            estimation, 170
        WLS fitting, 170
    residual, 177
    sample, 167
    spatiotemporal, 184
vector data, 9
vector data cube, 62
    examples, 99
vector to raster, 109
vector-to-raster, 10
vectors, 264

warning
    assumes attributes are constant
        over geometries of x, 50
warp
    rasters, 116
weighted Voronoi diagram, 191
weights
    inverse distance, 205
    listw, 204
    objects, 204
    row-standardised, 205
well-known binary, 33

well-known text, 30
WGS84, 27
WKB, 33
WKT, 30
WKT-2, 25, 26

write_mdim, 141

xts, 182

Zarr, 140

Index of functions

aggregate, 9, 81

class, 268
coord_sf, 127
count_components, 196

diameter, 207
dim, 269
distinct, 80
dnearneigh, 199

errorsarlm, 248

filter, 80
fit.variogram, 170

gabrielneigh, 198
gather, 80
gdal_utils, 98
geary.test, 215
glance_htest, 215
graph.adjacency, 196
group_by, 80

impacts, 253

joincount.multi, 212
joincount.test, 212

knearneigh, 200
knn2nb, 200

lagsarlm, 248
lisa_num_nbrs, 229
lisa_values, 229
lm.morantest, 214
local_multigeary, 229
localC_perm, 226
localG, 225
localmoran, 218

localmoran.sad, 222
localmoran_perm, 219

moran.mc, 215
moran.plot, 216
moran.test, 214
mutate, 80

n.comp.nb, 196, 198
nb2listw, 196, 204
nbdists, 197, 200
nblag, 207
nblag_cumul, 207
nest, 80

pivot_longer, 80
poly2nb, 194
predict.stars, 93

queen_weights, 229

read_mdim, 109
read_sf, 80
read_stars, 85
relativeneigh, 198
rename, 80
rowwise, 80

s2_closest_edges, 199
sacsarlm, 248
sample_frac, 80
sample_n, 80
select, 80
separate, 80
separate_rows, 80
sf_column, 76
sf_proj_info, 112
sf_project, 111
sf_use_s2, 195
shortest.paths, 207

slice, 80
soi.graph, 198
spautolm, 248
spread, 80
spweights.constants, 204
st_agr, 50, 76
st_apply, 96
st_area, 36
st_as_stars, 109
st_axis_order, 115
st_bbox, 76
st_boundary, 37
st_buffer, 37
st_cast, 37
st_centroid, 37
st_collection_extract, 37
st_contains, 36
st_contains_properly, 36
st_convex_hull, 37
st_covered_by, 36
st_covers, 36
st_crop, 88
st_crosses, 36
st_crs, 110
st_difference, 38
        n-ary, 38
st_dimension, 36
st_disjoint, 36
st_distance, 37
st_equals, 36
st_equals_exact, 36
st_geometry, 76
st_interpolate_aw, 82
st_intersection, 38
        n-ary, 38
st_intersects, 36
st_is, 34
st_is_empty, 34
st_is_longlat, 34
st_is_simple, 34
st_is_valid, 34
st_is_within_distance, 36
st_jitter, 37

st_join, 81
st_length, 36
st_line_merge, 37
st_make_grid, 82
st_make_valid, 37, 194
st_node, 37
st_overlaps, 36
st_point_on_surface, 37
st_polygonize, 37
st_precision, 39, 76
st_rasterize, 10
st_relate, 36, 37
st_segmentize, 37
st_set_agr, 76
st_set_bbox, 76
st_set_precision, 76
st_simplify, 37
st_split, 37
st_sym_difference, 38
st_touches, 36
st_transform, 37, 111
st_triangulate, 37
st_union, 38
        n-ary, 38
st_voronoi, 37
st_warp, 116
st_within, 36
st_wrap_dateline, 37
st_zm, 37
structure, 271
summarise, 80

transmute, 80
tri2nb, 197

ungroup, 80
union.nb, 207
unite, 80
unnest, 80

write.nb.gal, 196
write_mdim, 109
write_sf, 80

Milton Keynes UK
Ingram Content Group UK Ltd.
UKHW051535141024
449569UK00001B/44

9 781138 311183